What others are saying about this book:

"Technology acceptance, adoption and implementation are at the forefront of the today's valued REALTOR®. Since buyers and sellers are shown to be extremely web-savvy, this book promotes core real estate, web-centric technologies that further enable the committed REALTOR® to continue in well-exceeding clients' expectations."

—Joshua Sharfman, Ph.D.
Chief Technology Officer
California Association of Realtors®

"For real estate professionals, there is a sea of web-based technologies that one can adopt and implement. As an author of many technical books, this book is a refreshing piece that discusses technology from the perspective of the practitioner, the end-user. Real estate professionals looking to implement technologies to better engage business objectives should read this book as it gives purpose, reason and clarity to today's real estate technologies."

—George Reese
Chief Executive Officer & O'Reilly Author
Valtira Corp.

"Working with the Webographers organization has been an excellent opportunity to further our mission of helping real estate professionals discover greater time efficiencies, data accuracy, and professionalism. This valuable book explains how electronic forms software, online transaction management and other integrated technical solutions help streamline the workflow of the transaction as well as enhance the value of the REAL-TOR®. As we enter the era of the paperless transaction, the information in this text is on the need-to-know list of every professional in the real estate industry."

—Lisa Mihelcich
Chief Operating Officer
RE FormsNet/ZipForm

REAL ESTATE WEBOGRAPHER™:

Web Technology Handbook

REAL ESTATE WEBOGRAPHER™:

Web Technology Handbook

Marc Grayson

iUniverse, Inc.

New York Lincoln Shanghai

REAL ESTATE WEBOGRAPHER™: Web Technology Handbook

iUniverse books may be ordered through booksellers or by contacting:

iUniverse
2021 Pine Lake Road, Suite 100
Lincoln, NE 68512
www.iuniverse.com
1-800-Authors (1-800-288-4677)

ISBN-13: 978-0-595-39419-7 (pbk)
ISBN-13: 978-0-595-83816-5 (ebk)
ISBN-10: 0-595-39419-1 (pbk)
ISBN-10: 0-595-83816-2 (ebk)

Printed in the United States of America

Table of Contents

Advisory Group xvii

    Additional Acknowledgements xix

      Corporate Sponsor Personnel ....*xix*

      Real Estate Technology Advocates ....*xx*

Introduction xxiii

    Internet Users Today xxiv

    Webography xxv

    Next Step with the Internet xxvi

    Webographer vs. Web Designer xxvi

    Webographers.com xxvii

    How to Use This Book xxvii

    What's Inside xxix

PART ONE: *Welcome to the Web* 1

1. Embracing Technology 3

    Intended Audience 4

      Agents, Brokers, Support Staff *4*

      Independent Contractors *4*

    Technologies & Services Showcased 4

      Agent Websites *5*

      Single-Property Websites *6*

Virtual Tours *8*

Neighborhood Search *9*

CMAs and AVMs *10*

Electronic Forms *12*

Online Transaction Management *14*

Mobile Technologies *15*

Virtual Assistants *16*

Learning Laboratory™ **17**

2. Agents, Brokers & Support Staff **19**

Client Impressions **19**

Impress through Web-based Technologies **21**

Technology Deficiency **22**

Webographers.com Solution *23*

Structure of a Real Estate Agency **24**

Agent Process in Selling Properties *25*

Technology Roles within the Agency *25*

Role-based Technology *27*

3. Independent Contractors **29**

Independent Professionals **31**

Virtual Assistants *32*

TeamDoubleClick.com **35**

PART TWO: *Your Initial Web Presence* **39**

4. Webography **41**

The Webography Process **41**

Phase 1: Determine Your Requirements *42*

Phase 2: Select Required Technologies *44*

Phase 3: Select Service Providers *47*

Phase 4: Establish Seamless Appearance *51*

Webography Certifications **52**

5. Agent Website (The Main Presence) 53

    Internet Service Provider 53

    Traditional Hosting Companies 56

    Webographer-friendly Hosting Companies 58

        Control Panel *59*

        Templates *59*

        In-Browser, Web Page Editor *60*

        Inclusive Lead Generation *62*

        Domain Name *63*

        Signage *65*

        Business Email *66*

    RapidListings.com 68

PART THREE: *Connect Consumers through Internet Marketing* 73

6. Bring Traffic to an Agent Website 75

        Optimization for Search Engines *75*

    Home Town: Search Engine Optimization 77

    Nationwide: Search Engine Optimization 78

    Banner Ads 79

    Link Exchange Programs 80

7. Single-Property Websites 81

    One Property, One Website 81

    Residential Real Estate 83

        Advertising Single-property Websites *83*

        Real Estate Classifieds *84*

        Enhanced Advertising *85*

        Signage *86*

        "For-Sale-By-Owner" Argument *87*

        Assisted-Selling Facts *88*

        Property Management—"For Lease" *88*

Commercial Real Estate 89

    Office Space *90*

    Industrial Space *91*

    Vacant Land *92*

    Shopping Center *93*

    Multi-Family Listings *93*

    Retail-Commercial Listings *94*

    Hospitality Listings *95*

    Real Estate Classifieds *95*

    Enhanced Advertising *97*

AgencyLogic.com **98**

8. Virtual Tours **102**

Create a Virtual Tour Yourself **103**

    Overarching Practices *103*

    Commonly Used Practice *104*

    How to Shoot for Virtual Tours *104*

    Produce & Publish the Virtual Tour *107*

Copy to CD, floppy, email, etc **113**

Place Virtual Tours Everywhere **113**

    Agent Website *113*

    Single Property Websites *115*

    Online Classifieds *116*

Realtor.com and Virtual Tours **116**

    PicturePath™ program *117*

    Hometour360™ program *118*

RealBiz360.com **119**

PART FOUR: *Extend Your Web Presence* **123**

9. MLS IDX VOW ILD Technologies **125**

Multiple Listing Service (MLS) Overview **125**

Agent's Website & MLS Listings 127

   Internet Data eXchange (IDX) *127*

   Virtual Office Websites (VOW) *130*

   ILD to Replace VOW and IDX *130*

10. Neighborhood Search 132

  Lead-Generation 133

   Comsumer Websites *133*

  Neighborhood Search 134

  Lead-Generation—Agent's Web Presence 139

   Neighborhood Search on an Agent's Website *139*

  Neighborhood Search—Link Everywhere 140

   Agent Website *140*

   Single-property Websites *142*

   NeighborhoodScout.com *142*

11. CMA's and AVM's 147

  CMA Development Process 148

  Issues with the CMA from the MLS 149

   Market Solutions for the CMA *152*

  Automated Valuation Model (AVM) 153

  AVM vs. CMA 153

   AVMs in the Lending Industry *154*

   AVMs in the Consumer Market *154*

   AVM Use by Agents & Brokers *157*

   Provider—AVMs for Agents & Brokers *157*

  AgentAvm 158

   Back-office Reporting *161*

  AgentAvm—Link Everywhere 163

   Agent Website *163*

   Single-property Websites *164*

  AgentAvm.com 165

PART FIVE: *The "Paperless" Transaction* 167

12. Electronic Forms 169

    Alternatives to Electronic Forms Software 169

    Electronic Forms Software 170

        Form Libraries *172*

        Methods to Access Electronic Forms *172*

        Digital Signatures *176*

        MLS Integration *178*

    Online Transaction Management Perspective 179

    Utility Ordering 181

    ZipForm 183

13. Online Transaction Management 185

    The Need for OTM 187

    Who's Enabling Transaction Management Tools 188

    Benefits of OTM 189

        Broker/Owner *189*

        Agent *190*

        Peer Professionals *190*

        The Consumer *190*

    Features of OTM 191

        Task Tracking and Management *192*

        Digital Document Management *194*

        Participant Set-Up and Security *200*

        Communication, Notification and Logging *203*

        Service Ordering *207*

        Searching & Reporting *207*

    Transaction Coordinators 208

        Virtual Assistants *209*

    Personal & Corporate Branding 210

    OTM—Link Everywhere 211

    Agent Website *211*

    Single-property Websites *213*

  REBT.com *214*

PART SIX: *On-the-Go and in the Office* *215*

14. Mobile Technology *217*

  Mobile Devices *217*

    SmartPhones *218*

    Personal Digital Assistant (PDA) *218*

    Notebook PC *219*

  Email Notifications from Web-based Technologies *219*

  Mobile Applications Unleashed *220*

  iseemedia *220*

    iseerealty *221*

15. Additional Hardware & Software *225*

  Hardware *225*

    Personal Computer (PC) *225*

    Digital Camera *226*

    Camera Lens *227*

    Tripod/Panorama Head *228*

    Scanner *228*

  Software *229*

    Portable Document Format (PDF) Document *229*

    Digital Imaging Editor *229*

    Stock Photography *230*

    File Transfer Protocal (FTP) Software *232*

PART SEVEN: *The Future of Technology Adoption* *235*

16. Webographers.com *237*

  Technology Acceptance *238*

Online Training Experience 239
   Features Provided to Candidates *240*
Life after Certification 241
   Certification Renewal *242*
Road Ahead 242

PART EIGHT: *Appendices* 243
   APPENDIX A: Virtual Assistant Guide per State 245
   APPENDIX B: RapidListings MLS/IDX Coverage 297
   Index 315

Advisory Group

The REAL ESTATE WEBOGRAPHER™ Advisory Group is a select group of individuals providing peer-review of this published book and online course materials for the REAL ESTATE WEBOGRAPHER™ certification found at Webographers.com.

John Mathers

Director of Marketing
Company: RE FormsNet, LLC
Product: ZipForm®
Competency: Electronic Forms

Joshua Sharfman

Chief Executive Officer
Company: Real Estate Business Technologies, LLC
Product: RELAY™
Competency: Online Transaction Management

Steve Marques

V.P., Emerging Markets
Company: iseemedia Inc.
Product: iseerealty™
Competency: Mobile Technologies

Scott Reznicek

V.P., Strategic Initiatives
Company: eAppraiseIt, LLC

Product: AgentAVM™
Competency: Automated Valuation Model (AVM) Report

Gayle Buske

Chief Executive Officer
Company: Team Double-Click, Inc.
Product: Virtual Assistant Staffing
Competency: Virtual Assistants (VAs)

Andrew Couture

V.P., Business Development
Company: Location, Inc
Product: NeighborhoodScout®
Competency: Neighborhood Search

Philip Bliss

President & Chief Executive Officer
Company: x2idea Corp. (on-behalf of <u>RealBiz, LLC</u>)
Product: RealBiz360.com Tourbuilder
Competency: Virtual Tours

Stephen Fells

Chief Executive Officer
Company: Network Earth, Inc.
Product: AgencyLogic PowerSites™
Competency: Single-Property Websites

Jason Futch

Chief Executive Officer
Company: JCom Designs, Inc
Product: RapidListings.com
Competency: Agent Websites

Marc Grayson

President
Company: National Institute of Webographers, LLC
Product: REAL ESTATE WEBOGRAPHER™
Competency: Training, Assessment & Certification

George Reese

Chief Executive Officer
Company: Valtira Corp
Product: Simplicis Identity ™
Competency: Learning Laboratory™

Additional Acknowledgements

Corporate Sponsor Personnel

National Institute of Webographers, LLC
Tony Grantham
Davien Burnette

RE FormsNet, LLC (ZipForm®)
Lisa Mihelcich
Bill Tonnisen

Real Estate Business Technologies, LLC (*RELAY*™)
Howard Weinstein
Cassandra Davis

Location, Inc (*NeighborhoodScout*®)
Shannon Scott
Dale Lum

RapidListings.com
Jeremy Futch

RealBiz, LLC
> Sherry Marques

iseemedia,Inc
> Marsha Scharf
> Andy Opala

Network Earth, Inc (AgencyLogic.com)
> Dataigh Mullan

Valtira Corp
> Tom Ellingson
> Morgan Catlin

Real Estate Technology Advocates

e-Pro—Internet Crusade

National Institute of Webographers™ would like to thank and acknowledge the efforts of Internet Crusade and their technology training vessel of e-PRO, an Internet designation of the National Association of Realtors®. The efforts of Internet Crusade and the e-PRO designation look to advocate technology awareness of Realtors.

e-PRO designees have a solid understanding of Internet terminology and application for real estate objectives. Webographers.com expands on this foundation, having its candidates work with real technologies, hands-on, to further promote technology acceptance and adoption.

National Institute of Webographers™ feels that e-PRO should be taken by real estate professionals <u>before</u> undergoing the online training for the REAL ESTATE WEBOGRAPHER™ certification for those requiring basic understanding of the Internet and its application to real estate. Many competencies taught in e-PRO training are assumed "knowledge" of candidates undergoing training for the REAL ESTATE WEBOGRAPHER™ certification.

Center for Realtor® Technology

National Institute of Webographers would like to thank and acknowledge the efforts of Mark Lesswing and the Center for Reator® Technology (CRT) as the "technology advocate, implementation consultant and information resource". The CRT's development of applications, published surveys and white papers on real estate technology highlights why they care about technology acceptance and adoption by Realtors®.

Throughout this book and online course found at Webographers.com, training content is written in the "spirit" of the lessons and best practices expressed by the Center of Realtor® Technology (CRT). In addition, the National Institute of Webographers looks to showcase technologies that provide as *solutions* to the many reported issues expressed by the Realtor®-practitioner in the CRT's surveys.

Introduction

The real estate market has seen an *explosion* of activity in the last ten years due to the Internet. The Internet facilitates extensive activities, like the brokerage of real estate, than what was found in the past. The business methodology employed by real estate professionals has been completely altered due to the capabilities found online.

In terms of listing and advertising properties for-sale, hardcopy newspapers and real estate magazines were the medium of choice ten years ago. Potential buyers were limited to short descriptions, i.e. "4 bdr, 2.5 baths, deck…" Simply including a photo with these classified listings was an added "plus" due to the limited space of these advertising mediums.

Now, with the dawn of agent websites, single-property websites, and virtual tours, potential buyers are given a "real-life" view of a property from the comfort of their own home. This realism has allowed potential buyers to browse properties as if they were attending an "open house" in-person.

Neighborhood search and home valuation resources added to an agent's website has provided potential buyers and sellers value-added demographic and property information. Buyers and sellers are conducting such research on their own time, a trend commonly found with today's web-savvy consumer. Not only do these value-added resources keep site visitors on an agent's website longer, but provides "leads" directly to the agent whose hosting the website.

Electronic Forms and Online Transaction Management have provided an innovative way to allow the buyer and seller more "vision" and participation into the listing and closing process. Besides such web-enabled applications organizing and streamlining the transaction, buyers and sellers are more *aware* of the agent's listing and closing efforts. Before, such efforts were possibly unseen and under-valued. Now, these efforts can be mitigated and facilitated through the Internet; enabling a paperless transaction!

Mobility is the new business requirement amongst top-producing agents. The ability to conduct activities such as MLS searching and virtual tour viewing, more so than just email, is now possible from a wireless, mobile device.

Internet Users Today

Consumers of the real estate market are extremely web-savvy, where real estate professionals must "keep up" with this audience. Overall, the Internet has created an environment for consumers to easily complete tasks, like finding a home, which previously took more effort prior to the popularity of real estate found on the Internet.

Consumers already use the Internet to order airline tickets, pay bills, perform online banking, and more. In terms of real estate, investors are even purchasing properties without having to visit the property in-person. This convenience experienced by consumers is due in-part to the efforts of real estate professionals establishing a more effective web presence.

A February, 2004 telephone survey by Nielsen/NetRatings says about 204.3 million people or an estimated 75% of the population had Internet access at home[1]. Essentially, more people are online, and the numbers keep growing. With the increasing number of users online, what percentages of people use the Internet for purposes of buying a home?

> *According to the National Association of Realtors, 77 percent of all people looking for a new residence conduct a search on the Web.*[2]

Even though many users start their search for that perfect property online, many real estate brokerages and individual agents *still* remain to create and use an effective web presence. The National Institute of Webographers and REAL ESTATE WEBOGRAPHER™ professionals are committed to real estate agents, brokers and independent contractors establishing a marketable, effective, and user-friendly web presence for themselves or for their clients!

---

1. *Three out of four Americans have access to the Internet According to Nielson//Netratings.* Nielson//Netratings. March 18, 2004. http://direct.www.nielsen-netratings.com/pr/pr_040318.pdf.

2. *2005 National Association of Realtors® Profile of Home Buyers and Sellers.* National Association of Realtors.

Webography

As defined by the National Institute of Webographers, <u>Webography</u> is the art or practice of establishing a *seamless* web presence through the selection of various web-based applications, joined together through basic and applied web techniques.

Webography implies using out-of-the-box solutions for your web presence. When one considers business requirements like an agent website, virtual tours, electronic forms and online transaction management, is there *one*, all-inclusive service provider in existence of these technologies? The answer to that question is regretfully no.

An agent must establish customer relationships with various service providers, being mindful of how they can be joined together for seamless functionality, look, and appearance. After selection of these service providers, the agent or assistant applies basic web techniques (as described in this book), to establish a seamless web presence. Such efforts to establish a seamless web presence includes: selecting products "natively" compatible with each other, ensuring the same personal and corporate branding is applied across all web applications, selecting similar color schemes, and much more.

This process is an *evolution* from traditional web design to establish and maintain an agent's real estate, web presence. It empowers agents and their assistants the ability to self-serve their business objectives online the easy way. There are many providers of web-based applications for real estate, thus it may be a challenge for real estate professionals to choose the right services that best fits his or her endeavors and budget.

An agent must remember that establishing and maintaining a web presence is not just about technology. Based on the selected technologies and time-commitment to maintain each one, it may be concluded that an assistant is required to provide maintenance and support. Virtual assistants (VAs) are the growing trend for top-producing, real estate agents to focus on establishing new clients and making sales. VA's can provide that "man-power" to maintain an agent's web presence on their behalf.

Next Step with the Internet

What's the next step with the Internet if (a) users spend more time online, (b) establishing a web presence has been made easier, and (c) entities still remain to create an effective web presence? The answer is simple. The age has come for a "new" individual to establish a real estate web presence; whether for themselves, their company, or for clients in the real estate market.

A single individual can complete the entire process to establish a web presence when all real estate technologies are selected and documentation, content, and artwork for the web presence are provided. These individuals are known as REAL ESTATE WEBOGRAPHER™ professionals. These professionals possess demonstrated skills to not only create a web presence, but also have knowledge of the best real estate applications on the market that could encompass a web presence.

This book and the e-Learning experience found at Webographers.com, provides an extended overview of web technologies/services and how to implement them for yourself, your employer, or a given client. The web technologies and services are detailed from a vendor-neutral standpoint. However, corporate sponsors of the REAL ESTATE WEBOGRAPHER™ certification are showcased to provide as example, or real "case-study" to bring the lessons "home". The services highlighted are tailored for both the (1) do-it-yourself audience of brokers, agents, and support staff, in addition to the audience of (2) independent contractors, such as virtual assistants.

Webographer vs. Web Designer

Traditionally speaking, a web designer is one who creates a web-presence from "scratch". Such activities performed by a web designer may include: custom design of the front-end of the website, followed by back-end programming to make a website interactive.

A webographer on the other hand, uses pre-fabricated web sites and applications serviced by various technology platforms. A webographer then makes these web pages and applications *"their own"* through a few mouse clicks and strokes of a keyboard. Such basic configuration employed by a webographer, allows site users to transition from one application to the next in a seamless fashion. Ease in configuring a web presence with these providers allows for modification to the "look and feel" of these prefabricated web pages and applications, allows personal/corporate branding, all with no knowledge of HTML!

Webographers.com

From the experiences and shared best practices of real estate professionals, web professionals, and technology-related companies, comes the National Institute of Webographers. The company's primary service includes training, assessment and certification in real estate, web-based or web-enabled technologies. Certification available and achieved by candidates at www.Webographers.com includes:

- REAL ESTATE WEBOGRAPHER™

The REAL ESTATE WEBOGRAPHER™ certification is <u>not</u> a real estate license. Rather it denotes standards of comprehension in applied use of web applications for real estate objectives.

REAL ESTATE WEBOGRAPHER™ | **REW** ™

The certification marks above are owned by the National Institute of Webographers, LLC and are awarded to those who successfully complete initial and ongoing certification requirements.

Figure 1: NIW Certification Marks with Tagline

How to Use This Book

This book should be used as a handbook and guide for a REAL ESTATE WEBOGRAPHER™ professional. The primary audience for the REAL ESTATE WEBOGRAPHER™ certification includes (1) agents, brokers, & support staff, in addition to (2) independent contractors; those who assist real estate professionals establish and/or maintain a web presence.

Read each chapter, as the book introduces the REAL ESTATE WEBOGRAPHER™ certification, describes web applications and techniques specific to the REAL ESTATE WEBOGRAPHER™ professional, and describes a path to success as a REAL ESTATE WEBOGRAPHER™ professional. The book truly tells a story, step-by-step, on how to use the REAL ESTATE WEBOGRAPHER™ certification in a way that matches your interests, personality, and goals:

Identify your strengths and weaknesses with Internet applications. Understand and gauge your true level of expertise with Internet applications that encompass a real estate agent's web presence. The training and certification program for REAL ESTATE WEBOGRAPHER™ professionals denotes a "baseline" of comprehension in establishing and maintaining such a web presence. This book details applicable activities that can be performed by REAL ESTATE WEBOGRAPHER™ professionals, in consideration of that baseline.

Hardware and software requirements. Many of the software and hardware tools mentioned in this book you may already have at home or office. You have made the decision to focus as a REAL ESTATE WEBOGRAPHER™ professional. Let's determine what other tools are needed to succeed in establishing and maintaining a real estate, web presence.

Occupational Direction. Many of the certificants who have completed the REAL ESTATE WEBOGRAPHER™ certification have gained employment or propelled their careers within a real estate agency or company. Brokers, office managers and employers value the certification, as many positions in real estate agencies or staffing companies are reserved for REAL ESTATE WEBOGRAPHER™ professionals. Current real estate professionals, i.e. agents and brokers, view the certification as benchmark for being "web-savvy", to self-serve their business requirements on the Internet.

Entrepreneurial Interests. Many of the REAL ESTATE WEBOGRAPHER™ professionals nationwide choose to start a business revolving around the REAL ESTATE WEBOGRAPHER™ certification. This includes the audience of independent contractors, resellers, and virtual assistants.

On the other hand, agents and brokers are inherently entrepreneurs by trade. Although agents may work for a larger entity, success and income is driven by the individual. Agents must learn how to establish an effective web presence for themselves to be competitive in their local market.

Checklists, Worksheets, Resources, and More!

This book highlights various points of interest through "shaded" tables that should be noted by the reader. The various tables used throughout the book and their meanings are mentioned below:

Caution!

Denotes information in this book that provides as a warning.

Resources!

Denotes additional links to informational websites, books, publications or other highlighted literary references and resources.

Tips!

Denotes tips and best practices for a given topic.

Checklist!

Denotes a checklist of items that should be used for a acquiring a certain technology or performing a function.

Worksheet!

A blank worksheet that prompts the reader to plan a given procedure or set of actions.

Activity!

Describes a hands-on activity that can be performed at Webographers.com!

Footnotes
Detailed and descriptive footnotes are provided at the bottom of each page versus endnotes. Readers will note source information of a given reference for immediate lookup to further examine an article, industry study, survey, etc.

What's Inside

The book is in-concert with the online training provided at Webographers.com for the REAL ESTATE WEBOGRAPHER™ certification. This book is recommended reading for certification candidates at Webographers.com. In addition, this handbook contains day-to-day operational information on performing duties as a REAL ESTATE WEBOGRAPHER™ professional. After reading thoroughly through the material, you will gain the following information:

Identify the tools and resources that should be included in a web presence for real estate objectives. Based on the objectives of a broker, agent, staff or clients, it's important to know what applications/resources should be included in a real estate web presence; i.e., website, virtual tours, images, contact/lead forms, online transaction management, etc.

Marketing and advertising the REAL ESTATE WEBOGRAPHER™ *mark.* Consider the independent contractor, who assists others to establish a web presence. The REAL ESTATE WEBOGRAPHER™ certification marks not only implies technical-savvy, but displays a mentality that you are accessible and can provide one-on-one assistance in establishing and/or maintaining a web presence for others.

For brokers, agents, and support staff, it is already implied you are proficient in the principles surrounding the brokerage of real estate. The REAL ESTATE WBEOGRAPHER™ certification mark further adds the understanding to buyers & sellers you are "savvy" in web technologies that support those principles. In addition, the certification mark means you *use* web-based technology services to support your clients.

PART ONE:
Welcome to the Web

Chapter 1: Embracing Technology

Chapter 2: Agents, Brokers & Support Staff

Chapter 3: Independent Professionals

Embracing Technology

The brokerage of residential and commercial real estate is occurring in every town across North America. Agents and support staff, who are REAL ESTATE WEBOGRAPHER™ professionals, can enable a web presence to further facilitate the brokerage of these properties. Such web applications in real estate include: agent websites, single-property websites, interactive virtual tours, electronic forms, online transaction management, automated valuation model (AVM) reports, neighborhood search, and more. Web applications in real estate also appeal to potential buyers who may be out-of-the-area and cannot readily visit a property, or even their agent, in-person.

Webography is a process in examining your goals and requirements, review of web-based technologies that meet those requirements, selection of service providers, and integration of all selected technologies into one seamless web presence.

These 4 phases of Webography are detailed in Chapter 4. The process truly details how Webography is different than traditional web design. There are many do-it-yourself or self-serve technology providers that supply "pre-built", web-based applications. Such technology providers include templates and inclusive functionality that require "tweaking" to match your brand identity and business objectives. Sometimes, one provider may not meet all of your requirements. A REAL ESTATE WEBOGRAPHER™ professional understands how to best integrate technologies across various service providers into one seamless web presence.

Intended Audience

The intended audience of this book includes those who are not only knowledge-able in computers, but need direction in specific Internet-driven applications that meet real estate, web-related objectives. There are two types of individuals who commonly complete the REAL ESTATE WEBOGRAPHER™ certification: real estate professionals and independent contractors (i.e. virtual assistants).

Agents, Brokers, Support Staff

The REAL ESTATE WEBOGRAPHER™ certification inspires real estate pro-fessionals to create an effective web presence for themselves. Such professionals may be computer-savvy, but need hands-on training and guidance in which web applications related to the real estate industry they should use. Where should they invest their money to essentially generate leads, acquire new customers, advertise properties, and facilitate the closing of a property?

Brokers and agency owners may decide to have a dedicated, "in-house" staff member who oversees the web presence of the entire agency. Or, agents may self-serve their own personal web-presence, to be a stand-out within the agency or in-respect to other agents in their local area. Chapter 2 fully describes how real estate professionals leverage the REAL ESTATE WEBOGRAPHER™ certification for their current business.

Independent Contractors

REAL ESTATE WEBOGRAPHER™ professionals may act as an independent contractor or virtual assistant, providing a support role to establish and/or main-tain a web presence for others. Let's consider real estate professionals (i.e. agents, brokers) who want a web presence to better interact with clients, but do not have the time to build one (nor to maintain one). Chapter 3 describes in greater detail how independent contractors and virtual assistants play a role in the creation and maintenance of a web presence *for* the real estate professional.

Technologies & Services Showcased

This book describes the growing real estate certification of the REAL ESTATE WEBOGRAPHER™ professional. REAL ESTATE WEBOGRAPHER™ professionals can complete simple and straightforward, web-based objectives in the market of real estate by first determining online requirements. After acquisition of technologies that support those requirements, this book details how to add personal or corporate brand to these technologies, making it your own. The core real estate techniques and technologies, as shown below, may not be supported by one provider.

Core Real Estate Technologies/Services

- Agent Websites
- Single-Property Websites
- Virtual Tours
- MLS/IDX/VOW/ILD Applications
- Neighborhood Search Integration
- Comparable Market Analysis (CMA) reports
  Driven by Automated Valuation Modeled (AVM) technology
- Electronic Forms
- Online Transaction Management
- Mobile Technologies
- Virtual Assistants
  Administrators of Technology

How does one encompass all required technologies into one web presence? How does one keep a standard "look and feel" that is seamless across all applications for viewing by buyers, sellers and site-visitors? These questions are answered through the process of Webography; the art or practice of establishing a *seamless* web presence through the selection of web-based applications, joined together through basic and applied web techniques.

Agent Websites

Agent websites in support of the brokerage of real estate provides for full descriptions and enhanced visualization of properties. Some of these visual aides include digital photos, virtual tours, online brochure generation, and more! Agent websites in support of the real estate market can provide for "real-life" views of properties, rather than short, textual descriptions commonly found in print advertisements.

Corporate Sponsor
_____

RapidListings.com

Single-Property Websites

RapidListings.com provides full-service websites for agents. Through an easy-to-use control panel, websites can be tailored to meet the needs and personal interests of each agent. No knowledge of HTML is required. Each agent website comes with "informational" web pages like *Real Estate FAQs, Buyer/Seller Mistakes* and a *News Feed*. Agents can add MLS IDX feeds, web pages from other 3rd party subscriptions, or make their own, custom web pages!

_____

In the 2004 National Association of REALTORS®: Technology Impact Survey, it's estimated that 50% of Realtors have their own personal website. About 21% of those surveyed plan on starting their own website in the future.[3] Although real estate agents may be a part of a parent brokerage, agent success (and income) is driven by the individual. Agents today realize that a unique agent website is their "storefront" to current or potential buyers and sellers.

An agent website represents the good name of the agent, acting as their "voice" in making their all-inclusive services known, not too mention their current listings. Contact forms and other web page forms help to solicit "leads" for the agent directly from their website. Many website providers have an inclusive contact manager tool that may be integrated with today's popular CRM (contact relationship management) tools. Commonly, an agent will use their personal name for the domain name (URL), associated with their website (i.e. www.GregDugan.com).

Single-Property Websites

Single-property websites put the seller's interest back in the forefront. As a resource used by the seller's agent, single-property websites (i.e. www.123AnySt.com) establishes a full web presence for one (1) property. They've become the most powerful listing tool for real estate agents. Single-property websites include photos, property

_____

[3] *Personal Real Estate Websites.* 2004 National Association of REALTORS®: Technology Impact Survey, p 21.

descriptions, floor plans, links to virtual tours, custom neighborhood and area information, and much more.

Considering the URL of a property website is commonly the street address (i.e. www.123AnySt.com), it provides for a unique and easy-to-remember web address. Besides mention of the URL on literature such as a CMA, consider mediums where advertising space may be small, such as a newspaper ad or signage placed outside the home. Placement of a single-property's URL in a newspaper Ad, or on a "sign rider" placed with the property's signage, appeals to the web-savvy buyer.

Providers of single-property websites normally have a "point & click" interface for agents to establish a unique website for one property within minutes. Such a provider commonly features the ability to create custom web pages with any information you desire. All-in-all, a single-property website acts as a full-service website for one (1) property. Rather than paying a monthly subscription, such providers of single-property websites commonly have a pay-as-you-go, per website plan!

Corporate Sponsor

AgencyLogic.com

Single-Property Websites

AgencyLogic.com and its inclusive PowerSites™ service are the cutting-edge way to win listings by instantly giving every client's home its own showcase Website, personal Web address and high-impact marketing plan. In an age where the competition to win listings is fierce, PowerSites are the next generation of Internet-enabled marketing tools that enable you to consistently out-market and out-perform your competition. An agent can prepare a PowerSite for a listing in minutes.

The real estate agent's secret to winning more listings - give every property you list its own showcase on the Web! Winning new business is all about persuading your prospect that you're going to sell their home faster and for the best price. That means having the best marketing plan and using the best marketing tools. [4]

---

[4] *One Property, One Website, One Solution. Network Earth. Retrieved June 13, 2006.* http://www.agencylogic.com

Virtual Tours

Eighty four percent of home buyers say photos and detailed property descriptions are the most important features when searching online for homes—followed closely by virtual tours.[5] When home buyers search through online "classified" listings like Realtor.com, many site visitors will only view listings that include virtual tours. Virtual Tours have provided real estate agents the ability to provide an "open-house" of a property 24x7. Such tours have enabled consumers, especially out-of-town buyers, the ability to view properties as if they were there in-person.

Home buyers clearly recognize the value of virtual tours, with the "Virtual Tours First" display option clicked more than 120,000 times each day on REALTOR.com.[6] Virtual tours increase the chance of matching a property to the purchaser's requirements.

Corporate Sponsor

RealBiz360.com

Virtual Tours

RealBiz360.com provides a complete, online, virtual tour marketing solution designed specifically for the Real Estate professional – that allows unlimited tours for one affordable subscription. High-Definition Virtual Tours (HDVT) delivers an unparalleled experience to the real estate buyer, enabling users to zoom-in and see fine details within a Wide panoramic viewer. This self-serve virtual tour solution provides tour upload to Realtor.com as a PicturePath™ provider.

Realbiz360 uses photo-stitching technology stemming from iseeMedia's Photovista® product line, hailed as the industry leader in virtual tour creation for nearly a decade; continuously recommended to beginners in virtual tour creation. That power and ease-of-use is brought forth at Realbiz360.com for agents, brokers and their assistants to self-serve their virtual tour needs.

---

[5] 2004 National Association of REALTORS® Profile of Home Buyers and Sellers.

[6] Move, Inc (formerly Homestore, Inc.) June 2004 internal analysis of total unique searches vs. searches with virtual tour sort. http://resource.realtor.com/agent/VirtualTours.aspx. Obtained December 10, 2005

Virtual tours provide the opportunity for potential buyers and other agents to preview your property prior to an actual visit. They are becoming the prime property listing resource for home buyers. Home buyers prefer previewing homes via virtual tours before viewing the physical property. They provide an accurate representation of the property and decrease the time and expense in showing of properties by the agent.

Once a virtual tour is created, it is a "reusable" resource. A hosted virtual tour can be hyperlinked from properties posted on an agent's website, a single-property website, or from online classified listings like Yahoo! *Real Estate* and Realtor.com. A quality virtual tour, solutions provider will allow agents to "burn" or copy their virtual tours to CD or even email the virtual tour.

Neighborhood Search

Neighborhood search, incorporated into an agent's website, helps to establish a full service website for homebuyers and sellers. Not only is the buyer's search for the perfect home imperative, but so is the search for the perfect neighborhood.

Value-added resources such as neighborhood search found on an agent's website keep visitors on the website longer. Also, such a resource ignites electronic communication between buyers & sellers and the target agent, providing a "lead generation" tool within the agent's website.

No matter what the real estate climate, consumers will be particular about the neighborhood in which they select a property. As it's known that consumers are web-savvy and interested in performing their own market research, adding neighborhood search to an agent's website caters to those interests of consumers "browsing" market information. However, that convenience of allowing the consumer to browse on their own time, will produce "quality" conversation (a lead) between the consumer and the agent who's providing neighborhood search from their website (consumers reach out to the agent).

Once placed on an agent's or broker's website, no additional maintenance is required on behalf of the agent. Agents respond to client requests (sent to the agent's email inbox or drip marketing program) to see homes in their chosen neighborhoods or communities. Commonly, real estate consumers would have to pay for such market knowledge from neighborhood search websites. Rather, an agent establishes a subscription and then incorporates the neighborhood search from their website. Here, the agent "foots the bill" for a specified number of searches for a given month, all-the-while obtaining leads sent directly to the sponsoring agent.

A solid provider of neighborhood search, added to an agent's website, does not charge any referral fees or lead generation fees. Thus, subscribing real estate professionals can save thousands of dollars when compared to the high referral fees on real estate "lead" websites.

Corporate Sponsor

NeighborhoodScout.com is making it easier for real estate professionals to turn their websites into a full service real estate destination for home buyers and sellers, by unveiling a new subscription that integrates NeighborhoodScout® into their own websites; helping them generate exclusive real estate leads from their own website traffic.

The search engine covers all 61,000 neighborhoods in America. NeighborhoodScout allows agents to provide detailed street maps, median house values, neighborhood appreciation rates, exclusive school district ratings, FBI crime rates, and more, for every neighborhood in America, with one single application that goes on the agent's website.[7]

CMAs and AVMs

Comparable Market Analysis (CMA), are reports generated by agents and brokers to identify comparable properties that have sold in the local area in-respect to a target property. The CMA, helps to determine a property's value, given other comparable properties (also known as "comps") that have sold in recent history that are close in proximity and have a similar "size" and "shape."

---

[7] *Real Estate Agents and Brokers: Add NeighborhoodScout to your website today.*. Location, Inc. Retrieved December, 10 2005. http://www.neighborhoodscout.com/real-estate-leads.jsp. Reprinted with Permission.

Issues with CMA Generation

The efforts an assistant, agent, broker may go through to generate a CMA report, *worth providing to a potential seller*, can be extensive and cumbersome. Such efforts may be questioned by the agent to "win-the-listing", creating a seller's CMA, as such a report is short-lived.

The MLS commonly comes with tools for members to generate a CMA, however, an agent must commonly "export" that information to a word processing application for example, as the report is not yet "print-ready". Here the agent will reorganize the data displayed, attempt to create graphs and charts, add their agent picture, company logo, contact information and any other personal or corporate branding.

CMA Meets the AVM

The National Institute of Webographers, due to its own extensive research, has found the CMA-generation process to be time-intensive for today's real estate professional. This is true especially of the *Seller's CMA*, agents looking to "win-the-listing", where the CMA has a short shelf-life. Based on that research, a new product to produce a desired "CMA-looking" report has evolved. This product includes the use of Automated Valuation Model (AVM).

Automated Valuation Model (AVM) report is a resource used by the financial and lending industry for many years. An AVM provider utilizes data sources such as county & tax records and other proprietary databases to generate reports, which at first-glance, may look similar to a CMA. The purpose of an AVM is to provide comparable data and an estimated property value through scientific formulas versus human intuition.

A new product has evolved combining the best of CMAs and AVMs, for agent's personal use. This product combines the corporate branding of a CMA, but the scientific rigor in property valuation and comparable data of an AVM. This new report can be used as a replacement to generate a "Seller's CMA". For those busy agents who need a fast solution, this new report can act as the "Buyer's CMA" or act as additional, supplemental research.

Corporate Sponsor

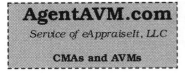
eAppraiseIt, LLC is the nation's leader in national appraisal management supporting the lending, appraisal and consumer industries. They have extensive experience with automated valuation products and have access to the industries most comprehensive data resources for AVM generation. These automated valuation models are now being used by real estate professionals nationwide and can be ordered directly from their website. This service, found at www.AgentAvm.com provides AVM technology to real estate agents and real estate brokers. This new service utilizes the most advanced AVM technology available in the industry today. It is designed to add value to the valuation process for real estate professionals and lenders alike.

With agent or broker accounts, real estate professionals can generate a full "CMA-looking" report along with: agent photo, contact information, personal/corporate branding, comparable sales information, property value and demographic data in a matter of seconds. In addition, AgentAvm.com allows for agents and brokers to place a banner on their own website so visitors can order a report for a specific property. This type of "site" order also provides the added benefit of capturing consumer contact information that can be used for future solicitation purposes.

Currently, many agents also use data sources like Public Information records of tax and county records to supplement MLS data. AVM technology uses such data sources in report generation. Thus AVM's used by agents can serve as a supplemental research tool for those who create "custom" CMAs.

Electronic Forms

Electronic transactions begin with the forms. Given listing contracts, sales contracts, disclosure statements, offers, counter-offers, etc., electronic form software allows for an organized and enhanced workflow in completing documentation within a transaction. Use of electronic form software provides an automated process of entering data one-time; with data then replicated across all the electronic forms related to the transaction.

For example, the property's street address, seller's contact information, etc., is entered one time. For agents using Word documents or PDF's as their preferred forms application, must manually enter in data related to the transaction numerous times. Such repetitive data entry, "opens the door" for data-entry error across all forms.

Providers of electronic forms solutions may provide "integration" with various technologies. In terms of consumer and property data, integration with the MLS allows for data to be imported directly from the MLS. Integration with online transaction management tools can provide transfer of forms to such an online repository and transactional interface for buyers, sellers and peer professionals.

Some providers of electronic form applications now provide online access to their electronic form libraries, or even access from a mobile device like a PDA. The online and mobile capabilities have given the on-the-go agent mobility; providing 24x7 access to edit and even email electronic forms. Thus, electronic forms and inclusive online/mobile access have truly extended the web presence of real estate professionals.

Corporate Sponsor

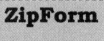

ZipForm®, known as the "Official Forms Software of the National Association of Realtors" dates its history back to 1991, as the first electronic real estate forms provider. At present, ZipForm has form libraries available for more than 370 associations and brokers. These form libraries are commonly navigated and completed using ZipForm*Desktop*, compatible with CRM tools like Microsoft Outlook®[8] and Top Producer®,[9] and the MLS via ZipForm*MLS-Connect.*

ZipForm*Online* has allowed REALTORS to go "beyond the office". There's no need to carry disks, or mistakenly work with the wrong version of an in-process contract. With ZipForm*Online*, you can use your laptop or any computer with an Internet connection, to securely access, print or email up-to-date forms to close the deal.[10]

---

[8] Outlook® is a registered trademark of Microsoft Corporation.

[9] Top Producer ® is a registered trademark of TOP PRODUCER® Systems, a Move, Inc company.

[10] *ZipFormOnline: GoBeyond the Office*. Electronic Marketing. Kit—ZipForm_GoBeyond.pdf. REFormsNet, LLC. Obtained November, 14 2005. Reprinted with permission.

Online Transaction Management

Online transaction management (OTM) has enabled brokers, agents, lenders, escrow & title companies, and buyers & sellers an interactive method to be intimately included in the closing process. Besides acting as a centralized repository of all documents/forms related to a transaction, OTM provides mobility for all parties to access documentation securely outside the office or home.

Each individual related to the transaction is given a unique login and password to access documents and forms of a given transaction. Permissions are applied to each user account; ensuring users are given access to specific documentation pertaining to them.

For Real Estate Professionals
For brokers, office managers, and especially risk management professionals, risks are mitigated knowing documentation is kept in a secure, online platform versus the backseat of an agent's car! With the volley of paperwork amongst the customer, agent, appraisers, title & escrow companies, OTM can enable a "paperless" transaction. OTM can ensure paperwork is not lost or misplaced, negating the potential for comprising the private information of the buyer or seller.

For Buyers & Sellers
OTM has provided the buyer/seller an insider's view of the efforts performed by agents and brokers during the closing of a property. Previously, such efforts involved in the closing of a property may have been overlooked by the buyer and seller. Thus OTM can provide the all-inclusive services consumers expect, services they can now see with their own eyes.

Corporate Sponsor

RELAY™, a service of Real Estate Business Technologies, LLC and located at www.REBT.com is the Transaction Management system designed by the REALTOR® practitioner. Hailed by industry leaders as a mission critical enhancement to the services provided by REALTORS® to homebuyers and sellers, RELAY™ allows the tracking and management of all information related to a real estate transaction from listing through closing; allows users to enable assistants, other agents or transaction facilitators to participate in the process; and features one-click integration with ZipForm® and WinForms Online®, the electronic forms software used by nearly 340,000 REALTORS® nationwide.[11]

Mobile Technologies

Notebook PC's (laptops), PDA's (Personal Digital Assistant), and SmartPhones have given mobility to the real estate professional. One can work on current tasks from any location, provide live demonstrations & examples to potential clients, or showcase work-performed to current clients.

Mobile technologies, as discussed in this book, go beyond standard operations like email, text messaging, and calendaring performed on a laptop or PDA. Imagine having virtually everything from your office desktop computer and having it available on a mobile device.

> *"Real estate agents were the early adopters of mobile phones. They are also the early adopters of PDAs. For an agent, being on the road meant they are either selling a property or acquiring a listing. The more time spent in the office means the less opportunity for making a sale. Now, the mobile phone and the PDA are converging, and again real estate agents are the early adopters."* [12]

Mobility is the new business requirement of real estate agents. With shared floortime at the brokerage, traveling to meet clients, conducting in-person visits, real estate agents desire to do more when *not* sitting in front of a personal computer.

Consider activities on your PDA such as: search and access MLS data, view virtual tours, review documents and much more. Such activities are paramount to conducting business on-the-go, while impressing your clients at in-person meetings with mobile technology.

In addition, many technologies like agent websites, single-property websites, and online transaction management, include email notification when "activity" has occurred on these platforms. As shown in this book, many technology providers allow for a secondary email address that is on-record in the agent's profile with the provider. This enables the agent to enter a "mobile" email address as their secondary email in their user profile; allowing the agent to respond timely to web-inquiries or activity from their mobile device.

---

[11] *The RELAY™ Advantage.* Real Estate Business Technologies, LLC. Retrieved December, 10 2005 http://www.rebt.com/RELAY.asp. Reprinted with permission.

[12] *iseerealty—Access MLS data, virtual tours, documents and more on your phone enable PDA or SmartPhone.* Iseemedia, Inc. http://www.iseemedia.com/main/products/isee-realtor. Retrieved December, 10 2005. Reprinted with permission.

Corporate Sponsor

iseemedia

Mobile Technologies

iseemedia Inc's, iseerealty™ is a mobile client server application using the iseemobility platform to deliver a must have application for all real estate agents. It provides four main areas of functionality: MLS data access; real estate applications; access to online Virtual Tours; and access to online documents. The application is designed to provide the agent with a strong complement of sales, marketing, communications and customer management tools to make the mobile device as effective as their desktop computer at the office or home.

iseemedia has developed a Real Estate virtual tour creation and publishing portal with partner RealBiz360.com. iseemedia's extensive research and development has reached the real estate market place via RealBiz360.com[13]

Virtual Assistants

All technology supporting real estate is invaluable, unless you have the time dedicated to maintain and use it. Real estate agents are natively busy people with returning phone calls, maintaining open house schedules, completing documentation related to the transaction, taking photos of properties, etc.

A virtual assistant (VA) is an independent contractor who provides an array of administrative services for the real estate professional. VA's may not necessarily reside in the same local vicinity as the agent. In this case the communication is transacted via email, telephone, fax, instant messenger, or a combination thereof.

VA's and Technology

Agents should never turn away business because they have become too busy. The solution resides in the use of virtual assistants (VA's) and independent contractors. From a technology standpoint, VA's can facilitate or coordinate documents

---

[13] *iseerealty—Access MLS data, virtual tours, documents and more on your phone enable PDA or SmartPhone.* Iseemedia, Inc. http://www.iseemedia.com/main/products/iseerealtor. Retrieved December, 10 2005. Reprinted with permission.

in a transaction, update an agent's website, generate and produce CMAs, and provide overall technology support.

VA's can perform administrative tasks such as: enter clients' leads into a CRM (contact relationship management) program, return phone calls on behalf of the agent, facilitate the paperwork involved in a transaction, set/cancel appointments, etc. VA's allow top-producing agents to focus on gaining new clients, closing more deals, and focusing on "face-time" with clients.

Corporate Sponsor

Team Double-Click

Virtual Assistants

Team Double-Click, Inc is the nation's leader in virtual assistant staffing for the real estate professional. With thousands of screened VA's on staff who are skilled in a variety of real estate competencies, explains why 75% of their client-base includes real estate professionals. Team Double-Click is the provider of the Team Contractor Real Estate (TCRE) certification for VA's. This unparalleled training experience ensures that VA's are competent, technically-savvy and know how to best support agents and brokers. With growing alignments with technology providers, Team Double-Click has specialized learning tracks for VAs, such as the Team Contractor Transaction Coordinator (TCTC) which highlights the RELAY™ transaction management system.

Learning Laboratory™

The online training and educational experience for the REAL ESTATE WEBO-GRAPHER™ certification, found exclusively at Webographers.com, is enhanced through the showcase of "live" real estate technologies at its Learning Laboratory™.[14] These technologies, offered in-part by the fore-mentioned Corporate Sponsors, provides demonstrative use of the real estate industry's foremost applications on the market. Such an approach in "technology awareness" training has never before been seen until now!

---

[14] Learning Laboratory™ is a trademark of the National Institute of Webographers, LLC.

The National Institute of Webographers understands that technology may be confusing to real estate professionals. As it stands, there are numerous service providers of real estate technology. Today's technologies are provided by the broker or as a part of MLS/Association member dues. Webographers.com provides its candidates demo access to today's most popular technologies to understand their business value, and most importantly, provide as a case-study to the learned processes, procedures and best practices.

At Webographers.com, each technology "competency" is complemented by an exemplary product currently on the market to act as a live case-study. Candidates complete assignments for a specific competency using one product. From a grading and assessment standpoint, using one product eliminates "bias" in grading and assessing candidates' work-performed.

Assignments and activities performed by candidates with a sponsored product at Webographers.com, look to be vendor-neutral. The purpose is to ensure training and comprehension on processes, procedures and best practices within a competency. For the candidate, it allows them to become comfortable with a product they may choose to purchase after training and use in business.

Candidates of the REAL ESTATE WEBOGRAPHER™ certification at Webographers.com utilize can one (1) <u>username</u>, <u>password</u>, and <u>candidate email</u> address to access all technologies across all Corporate Sponsors' platforms. Such capabilities allow the candidates "trial" accounts of all showcased technologies for the purposes of training, assessment, and certification.

Corporate Sponsor

Valtira

Learning Laboratory™

Valtira, Inc is the proud sponsor of the National Institute of Webographers, LLC's <u>Learning Laboratory</u>™. As powered by Valtira technology, candidates can have one universal username, password, and candidate email address during the training period to access an array of web-base, real estate technologies. The technology, provided in-part by Valtira, has brought all showcased technologies under one "umbrella", benchmarking Webographers.com as the most innovative, educational provider for real estate professionals.

Agents, Brokers & Support Staff

It is understood that real estate professionals must embrace technology that supports real estate objectives. Why? Clients and potential clients expect it! Customers want the ability to use the Internet (websites and email) when working with real estate professionals. Consider first-time home buyers. This group of individuals, possibly in their twenties, is Internet-savvy. They already perform many of their personal transactions online, with expectations of real estate companies and professionals to provide the same level Internet-capabilities.

Internet-savvy users already perform daily functions like: online banking, paying bills, planning vacations, and ordering gifts or merchandise online. The REAL ESTATE WEBOGRAPHER™ certification, achieved by agents, brokers and support staff, establishes technology awareness and hands-on expertise to meet the needs of such clients.

When a person sells property through real estate agent, there are many assumptions and expectations of what that listing agent and parent brokerage will provide. This chapter details how real estate professionals can not only meet, but *exceed* client's assumptions and expectations through web-based technologies.

Client Impressions

In many cases, sellers truly don't understand or know all the efforts performed by an agent in selling, and more importantly, the closing of a property. Why is this?

Many of the activities in the selling and closing of a property, customers simply don't see. However, the listing process with the MLS, agent website, and virtual tours *is* visible. Many agents do want a better way to show their sellers & buyers what value they bring to the "table".[15]

Value of Agents

Traditional expectations of a quality, real estate agent from the seller perspective may include:

- Help to determine the property owners selling power.
- Has many resources to assist buyers in finding your property:
    - Multiple Listing Service (MLS)

    - Newspaper listings

    - Listings in free magazines found in grocery/convenient stores.

    - Agent / Agency Website

    - Online Real Estate Classifieds

- A guide through closing process and final paperwork
- Provides current information on the local marketplace: current comparable pricing, available financing, neighborhood information, etc.
- Advice on when and how to advertise a seller's property.
- Provides recommendations for upgrades (fixes) to a property and methods to best stage the property for an "Open House".
- Organizes an "Open House" schedule.
- Helps evaluate buyer's proposals.
- Helps with the close of the property.

Web-based technologies can facilitate many of the functions above, providing a "visual" so that buyers and sellers can see all the valiant efforts of the agent. Providing the customer those visuals to see all efforts of the agent will provide referral and repeat business for the agent and parent brokerage.

---

[15] 2005 REALTOR Technology Efficiency Study. (February 23, 2004), WAV Group in conjunction with Center for REALTOR® Technology (CRT). p28. http://www.Realtor.org/CRT

Impress through Web-based Technologies

In terms of the seller, what are some of the activities they *can* visually see? Sellers cannot see first-hand, the hours of painstakingly completing paperwork, making copies, organizing open-houses, etc. However, they can see what efforts are put forth when those efforts are displayed online!

Online advertising of property listings
Real estate agents should proudly display all advertising efforts to the seller. An agent can show properties listed on 3rd party websites like Realtor.com, their agent website, and in other various online classified outlets. An agent can go as far as establishing a website dedicated to one property (single-property websites).

Virtual-tours created to provide a 24x7, "open house" of their property, have become "standard-issue" in the eyes of the seller. They search Realtor.com and agent websites. They as consumers know and understand the capabilities of web-savvy agents. Sellers will likely show their friends and family the work performed by agents using technology, providing additional word-of-mouth marketing of this agent and their parent brokerage. In addition, the seller has immediate gratification in using an agent to sell the property versus attempting on their own.

Value-Added Resources—an Agent's Web Presence
There are many "add-on" technologies an agent or agency can make inclusive of their web presence. These add-ons include "tools" to gather leads of potential buyers and sellers, or provide value-added resources. Examples of such technologies include Neighborhood Search, Automated Valuation Model (AVM) reports with the agent's personal branding, additional information on listed properties, etc. These value-added resources incorporated to a web presence can help to acquire "lead" information, for immediate follow-up by real estate professionals.

Electronic Forms & Online Transaction Management (OTM)
Electronic transactions begin with the forms. Besides providing an automated process to completing forms related to a transaction, it implies forms can be emailed to buyers and sellers since the forms are natively in electronic format. Buyers and sellers having work and personal email addresses are attracted to having transactional documentation communicated via email.

Online transaction management (OTM) provides a web-enabled interface with the transaction owner, sponsoring agent, their client, opposing agent and their client, lenders, appraisers, and escrow/title all through the life of the transaction. Such a resource not only provides a safe, central repository of relevant documentation, but provides for prompted electronic dialogue between agent, customer and other real estate professionals. Permissions can be established to provide customers limited access to the online transaction for a given property. This keeps customers informed and allows them to *see* the agents efforts involved with the closing of a home.

Web-based technology efforts provided on the behalf of the buyer & seller are something that they can visually see. These efforts explicitly demonstrate why the agent is involved in the selling process and is worthy of commission.

Technology Deficiency

Why even expose, train and teach technology to real estate professionals? Studies done by the Center for REALTOR® Technology (CRT) to examine technology use by REALTORS, gives credence to such training and certification. One study examined how technology is used for lead generation, selling and buying of real estate. The study is detailed in the 2005 REALTOR Technology Efficiency Study, published in February of that same year.[16] While many REALTORS® are tech-savvy, many still remain to embrace technology. Here are some points from the study that support that claim:

- *Age Gap*. The study implies that many customers are younger than agents. This rift has caused a "technology-gap".

- *Intimidation*. Several agents in the study say that technology intimidates them.

- *Awareness*—Brokers in the study said they were not aware of technologies that support real estate activities.

- *Assistants*—many agents have hired tech-savvy assistants due to the agent's lack of technical knowledge.

Caution!

Many web-savvy buyers and sellers are approaching agents with more market information than what the agents have!

---

[16] 2005 REALTOR Technology Efficiency Study. (2004, February 23), WAV Group in conjunction with Center for REALTOR® Technology (CRT). p29-30. http://www.Realtor.org/CRT

Sometimes consumers may acquire more information, or are more knowledgeable on some market specifics than the real estate professional! For example, buyers and sellers can go to the websites of many lending and financial institutions to acquire market data, such as property valuation estimates. Here, they can obtain an estimate of a property's value through an Automated Valuation Model (AVM) type-of-report. Although a scientific estimate and not of an agent's intuition, customers are readily accessing market data by their own means.

Potentially, the agent's lack-of acceptance and use of web-based technologies may have buyers and sellers finding the agent's processes outdated. There are many web-based technologies in existence to support the for-sale-by-owner (FSBO) for example. To combat the "do-it-yourself" mentality of FSBO's who are enticed by the Internet; real estate professionals must make web-based technologies a part of their business methodology and daily practice.

National Institute of Webographers is committed to real estate professionals and their support staff to accept and adopt web-based technologies. Their mission is to ensure professionals in real estate can meet the requirements and assumptions of their web-savvy customers. This mission is expressed through the online training for the REAL ESTATE WEBOGRAPHER™ certification found exclusively at www.Webographers.com.

Webographers.com Solution

The training provided at Webographers.com not only provides real estate professionals an overview of existing real estate technologies, <u>but allows them to "play" with such technologies</u>. It's this type of environment that alleviates intimidation of technology, as the listed web applications are showcased in a non-obtrusive, educational setting.

Technology Adoption
In addition, the applications showcased during training are provided by REAL companies, where the technology can be used for commercial use *after* training. Finally, discussion forums found at Webographers.com are a "hot-bed" of communication on topics of real estate applications and technologies among professionals across North America. These forums provide a network of real estate professionals sharing an array of best practices on various real estate technologies.

Given an agent's objectives, let's review the objectives of a given agency, to later examine how technology plays a pivotal role.

Structure of a Real Estate Agency

As an example, consider **AGENCY ABC**. This real estate agency deals primarily with residential real estate, providing services for both the buyer and seller. There are different types of personnel that a REAL ESTATE WEBOGRAPHER™ professional should be aware of in a real estate agency. Let's say the personnel that work at AGENCY ABC are:

1. Broker/Office Managers (2)
2. Agents—In the field (35)
3. Advertising Coordinator (1)
4. Transaction Coordinator (1)
5. Listing Coordinator (1)

In this agency, the brokers and agents all happen to be REALTORS®.[17] They are licensed professionals in the real estate industry and are members of the National Association of Realtors (NAR). The transaction and listing coordinators have experience, training, certifications and/or licenses to perform their support roles within the agency.

Broker/Office Managers. These individuals oversee day-to-day activities at the agency. They supervise the sales agents and support staff employees. They ensure training of all agents, and keep abreast of current laws and trends applicable to their region.

Agents. Both new and established agents are the "human link" between the agency and the buyer/seller. They perform many of the value-added functions with customers: phone calls, email communications, open-house visits, etc.

Support staff. The Listing Coordinator ensures that each new property is established as an "Agency Listed Property" and is entered in the local MLS. The Advertising coordinator is the lead support person in ensuring all listings of the agency reach the required outlets, both print and online. The Transaction Coordinator oversees all documents related to transactions from contract-to-close.

---

[17] REALTORS® is a federally registered collective membership mark which identifies a real estate professional who is a Member of the NATIONAL ASSOCIATION OF REALTORS® and subscribes to its strict Code of Ethics.

Let's now discuss the agent process in selling properties, to later express how technologies play a role in such real estate activity.

Agent Process in Selling Properties

So let's explain the selling/buying process of real estate when an agent is utilized by the consumer. A seller approaches a real estate agent from AGENCY ABC and signs a listing agreement to work with that agent (listing agent) and parent agency. The real estate agent or office assistant enters the listing in the MLS and advertises the property in the newspapers, real estate magazines, online, etc. The real estate agent or other agents within that agency, work in the field to get that home sold. However once that property is in the MLS, any real estate agent from any agency can sell that home.

Let's say commission on the property is 6%. If the agent from AGENCY ABC sells the home, that agency gets all the commission, splitting the commission however decided between the brokers and listing agent. Consider the event that an agent from a different agency, e.g. AGENCY XYZ, sells the home. The commission is then split 2 ways between the two agencies, each receiving 3%, with each agency dividing their 3% between those in the agency.

As you may conclude from the "agent process" scenario, the agent that first signs a client, wants to sell that client's property. In the scenario stated above, the agent wants all the commission from the sale of the home to come back to AGENCY ABC. To achieve this, that seller's agent must surround their name with that property in every advertising outlet they're using. Secondly, they must consider all mechanisms to advertise their properties to potential buyers to include the agent's or agency's brand name. Web capabilities utilized by a REAL ESTATE WEBOGRAPHER™ professional can do just that.

Technology Roles within the Agency

There are many roles that are played within a real estate agency. As technology has grown, more people in the agency must become comfortable working with technology. There are many manual processes that can be assisted through technology, and customers are now demanding such web technologies throughout the buying and selling process.

Brokers/Office Managers

In terms of technology, brokers/office managers play the primary role in establishing applications utilized by the agency as a whole. Yes, there are some applications that agents can purchase for themselves, but brokers play a part in establishing and procuring technology that supports the *entire* agency.

Resources!

Many web-savvy buyers and sellers are approaching agents with more market information than what the agents have!

- The Center for Realtor® Technology showcased at www.Realtor.org /CRT, commonly describes technologies reviewed for Realtors®.

- Clareity Consulting at www.CallClareity.com, contain research publications and mid-year reports about real estate specific technologies.

- Forums at www.Webographers.com are a hotbed of discussion on latest technologies acquired and implemented by Brokers across the North America.

Brokers can research online, network with their peers, or look toward their local and state associations and be an advocate to make certain technologies available as a part of member dues. Overall, they must also take a pro-active approach to acquire technology that propels the agency to provide the best service to their customers.

Agents

Agents should invest their time in examining technology that best supports their customers. Agents have the ability and obligation to highlight deficiencies in technology that supports agency customers, followed by recommendations to their broker. Why should such input come from agents? Agents are the ones that have first-hand contact with customers. It has been shown that REALTORS® spend 50% of their time on value-added activities like direct client contact, sending emails, taking phone calls or meeting with clients.[18] Thus agents are constantly in direct contact with buyers & sellers and best understand their needs, especially needs that can be met through web-based technologies.

Secondly, real estate is driven by the commission of the agent. Agents are inherently entrepreneurs. Agents must implement web-based technologies for themselves and establish their own web presence. This "do-it-yourself" mentality has created many top-producers who understand the value and timed-saved in using a proficient web presence, not to mention the "wow-factor" experienced by consumers.

Support Staff

Whether referring to listing coordinators, advertising coordinators, transaction coordinators, or office specialists, these individuals may be the most important people in implementing and maintaining technology for the agency. This is especially true if an agency supports many agents or is highly active in the brokerage of properties. Through technology, especially web-based applications, support staff can streamline many processes that may keep agents from performing value-added functions with their clients. Support staff working with technology can free up agents to better perform duties like direct client contact, sending emails, taking phone calls or meeting with clients.

For those top-producing agents who have their own web presence, maintenance can be time consuming. Adding and deleting listings to agent websites can be time-consuming for example. Personal assistants like virtual assistants (VAs) can type property descriptions on agent websites, upload property images, hyperlink virtual tours to the listings and more. In addition, virtual assistants can perform this same function on other advertising outlets, like Realtor.com, Yahoo! *Real Estate*, and other local online classifieds.

Role-based Technology

Many service providers have "role-based", user accounts for real estate technologies like: online transaction management, forms software providers, agency websites, etc. This means that brokers, agents and support staff can each have a personal user account, with various levels of permission, to conduct "role-based" activities within that corporate account. For example, a provider of agency websites may have user account for a broker or agent to establish the account and maintain the level of service purchased.

---

[18] 2005 REALTOR Technology Efficiency Study. (2004, February 23), WAV Group in conjunction with Center for REALTOR® Technology (CRT). p29. http://www.Realtor.org/CRT

User account for an assistant may allow them to maintain their specific web pages, forms, etc, without access to credit card information on file and other private account information. These types of role-based accounts are also effective in using independent contractors or virtual assistants, providing them restricted access to perform activities within the agent's account as required.

Online transaction management (OTM) systems are great examples of web-based applications that support role-based, user accounts for the buyer, seller, respective agents, title/escrow, mortgage/lender, inspectors, and other professionals.

Chapter 3

Independent Contractors

Many real estate professionals desire a web presence that fulfills their business objectives and keeps up with their web-savvy customers. However, some professionals do not have the time to create one, nor maintain one. Agency employees, who are REAL ESTATE WEBOGRAPHER™ professionals, may perform the duties described in this chapter in establishing or maintaining a web presence. Or, an independent contractor, such as a Virtual Assistant who is a REAL ESTATE WEBOGRAPHER™ professional, may be acquired to further support the web technology needs of the agency.

Utilizing an independent contractor may be desired for establishing a web presence. Or, it may be required to have such a person also provide on-going maintenance and long-term support. As it may, maintaining a web presence may require only impromptu updates or modifications that do not require a steady employee. Whatever is decided, the individual should be a REAL ESTATE WEBOGRAPHER™ professional to meet the required web objectives of the agent or broker.

Why is there a current need for independent contractors who are REAL ESTATE WEBOGRAPHER™ professionals? Why would real estate professionals need such outside assistance in establishing a web presence? Below are four main reasons:

1. **Basic Web Services Needed**
    Individual agents or whole agencies requiring a web presence for straight-forward objectives, normally want to avoid major web-design/IT companies for many reasons. One reason includes the feeling that the project is more of do-it-yourself or in-house endeavor. Also, overhead costs in dealing with a large web design company are usually included.

2. **Assistance with Existing Technologies**

Agents commonly ask, "How do I get started and build a web presence? What vendor(s) do I go with? What's the return-on-investment (ROI) of going online"? A REAL ESTATE WEBOGRAPHER™ professional acting as an independent contractor can guide such a client: help select the right web technology vendor(s), provide initial setup assistance, and provide basic training on how the client can self maintain the web presence. Or a REAL ESTATE WEBOGRAPHER™ professional acting as an independent contractor may also provide maintenance to a web site and inclusive technologies long-term.

3. **One-on-One Interaction**

A REAL ESTATE WEBOGRAPHER™ professional acting as an independent contractor has the ability to interact locally in their town (or virtually across state lines), providing administrative support and building customer confidence in working one-on-one with a real estate professional. The idea of one individual providing all-inclusive service to a client is commonly seen in other industries. Examples of such professionals in other industries include CERTIFIED FINANCIAL PLANNER™[19] professionals and REALTORS®[20]. These professionals carry certification marks or collective membership marks, respectively, which are protected by trademark laws. They not only provide services to clients in their given industry, but adhere to a code of ethics required by the owner of the trademarks. Most importantly, these professionals are accessible and provide one-on-one attention.

4. **Commonality of the Target Market**

Real estate brokering will always be occurring in every town. New agents will continuously emerge. Brokers and agents will always be looking to provide greater support to customers through technology, as technology is constantly evolving. Thus REAL ESTATE WEBOGRAPHER™ professionals, who are independent contractors, have the ability to be suc-

---

[19] Certified Financial Planner Board of Standards Inc. owns the certification marks CFP®, CERTIFIED FINANCIAL PLANNER™ and federally registered CFP (with flame logo), which it awards to individuals who successfully complete initial and ongoing certification requirements.

[20] REALTORS® is a federally registered collective membership mark which identifies real estate professionals who are Members of the NATIONAL ASSOCIATION OF REALTORS® and subscribes to its strict Code of Ethics.

cessful in their own town (or connected beyond their state lines virtually) because the need for web support in real estate is widely needed.

Independent Professionals

A REAL ESTATE WEBOGRAPHER™ professional acting as an independent contractor completes web projects with their own resources (i.e. personal computer) and on their own time. Described throughout this book are technologies that encompass a web presence for real estate agents. Many of these technologies have high potential for independent professionals to provide setup and long-term administrative support to agents who *purchase* and *use* these technologies. One-on-one support provided to agents, on an "as-needed" basis, weekly or monthly schedule, is the key to agents becoming top-producers.

Providing Occupational Direction

After the dot-com era, there exist about 10.5 million IT workers, in the US alone, who are competing for jobs that have rising qualification requirements, as reported by the Information Technology Association of America. ITAA is an Arlington, Virginia-based trade group representing technology companies.[21] With such dramatic numbers of IT people competing for limited job openings, the time is right for a professional certification that inspires "work-for-yourself" potential.

Reseller of Products & Services
An independent contractor may act as a "Reseller" or "Affiliate" of preferred applications for real estate professionals. Here a REAL ESTATE WEBOGRA-PHER™ professional can sell services to clients under their own brand name, the brand of the service provider, or co-branded interface. As a reseller or affiliate, the independent contractor earns a commission for sales generated. Many of the Corporate Sponsors of the REAL ESTATE WEBOGRAPHER™ certification provide reseller or affiliate programs to enable the independent contractors to offer their services, along with a nice incentive such as discounts.

---

[21] Associated Press. Carpentor Dave. (2004, December 15). *Skipping college paid off for some teen techies, not al.* Retrieved December, 16 2004 from http://www.nctimes.com/articles/2004/12/15/special_reports/science_technology/21_21_1712_14_04.txt

Webographers

The term "webographer" rings synonymous to the likes of a photographer, videographer and other potential freelance professionals. REAL ESTATE WEBOGRAPHER™ professionals acting as freelance professionals may be a reseller as stated above, but also provide the inclusive service of consultation, setup, deployment, and maintenance of a complete web presence.

Webographer is a generic phrase. However, the REAL ESTATE WEBOGRAPHER™ mark denotes standards in applied web techniques and adherence to Commitment Document as set forth by the National Institute of Webographers, LLC (Webographers.com); the owner of the REAL ESTATE WEBOGRAPHER™ certification mark.

This type of freelance professional moderates the entire Webography process as mentioned in Chapter 4. They gather requirements, translate the requirements into common technologies, make recommendations on suggested service providers, and implement all technologies purchased by the client into one seamless web presence.

Virtual Assistants

Many REAL ESTATE WEBOGRAPHER™ professionals are virtual assistants. Likely, they do not hold licensure in the brokerage of real estate, but are trained to assist licensed professionals with day-to-day administrative support and maintenance of a web presence. Many noted authors have stated for a real estate agent to be top-producer, they must have an assistant so they can focus on procedures that bring in money and new clients. Below are some overarching items a VA can do. Later in this chapter, is an all-inclusive list of what a Team Double-Click, virtual assistants can perform to provide as a case-study.

> *Communication.* Consider a website that has a simple form or contact email address for a seller to inquire about the services of an agency or agent. Clients now-a-days find response times over 12 hours to be unacceptable.[22] VA's can filter and organize contacts to ensure an agent or agency doesn't miss any leads.

---

[22] 2005 REALTOR Technology Efficiency Study. (2004, February 23), WAV Group in conjunction with Center for REALTOR® Technology (CRT). p30. http://www.Realtor.org/CRT

- VA's can add the person as "New Prospect" to the agent's contact relationship management (CRM) tool. VA's can further add an item to agents schedule to "call" prospect.

Virtual Tours & Still Photos. Agents, brokers, and agencies should strive to have an inclusive virtual tour with every property listing. VA's can prepare and stitch photos for virtual tour publishing. In addition, a VA can organize the tour and type property details associated with the tour.

Marketing Materials. Some providers of agency websites provide the ability to auto-generate a brochure of a property listed on the website. Sometimes the agent may require a brochure with specific data.

- A VA can help generate a brochure on a property that meets the requirements of an agent or agency.
- A VA, if they provide contact management services to an agent or agency, may prepare a mailing (or emailing) group that should receive such a flyer.

  Or, a potential seller may want to learn what their home is currently valued. Such an inquiry is a "lead" for the potential of acting as the listing agent of that seller.

- A VA can prepare a CMA that includes information retrieved from an online system, to ensure clients receive current sales-history in their local area or property valuation information in a timely and efficient manner

Website Maintenance. Real estate in itself is a dynamic industry. Properties are constantly added and deleted from the marketplace. Websites in support of real estate must be as equally dynamic. This revolving door of activity may become overwhelming to an agent self-serving their web presence.

- VA's can add property images and virtual tours to the agency website. Links to the published tours can be added to various online classifieds and MLS listings.
- VA's can keep track and delete listing information when instructed to do so.

Graphic Design. Many of the technologies that encompass a website are template-driven. However, custom images and artwork can be added to convey a brand image for the agent or agency.

- VA's can assist in designing a "top banner" for an agent's or agency's website. This can entail the incorporation of a company logo, agent's photograph, and artwork related to real estate.

- VA's can design the "top-banner" of virtual tour displays. Commonly, a provider of self-serve virtual solutions will provide templates. Many-of-times, these templates have an area where custom artwork can be added to create a brand-image.

- VA's can develop the letterhead on reports generated for clients and any other print materials. This can be extended to include email messages that require a brand image on every message.

Transaction Coordination. Given an online transaction management tool for efficiency in the transaction process, virtual assistants can act as Transaction Coordinators, mitigating this process on behalf of the client agent or brokerage.

Caution!

Be sure to check local and state governments on the legalities of desired functions of virtual assistants. Some functions and capabilities must have specific certification or state licensure to perform the duties. *Assume the VA is not licensed.* As provided by Team Double-Click, Inc. Appendix A of this book is an overview of what licensed and non-licensed personnel can do in many of the 50 states.

Virtual assistant staffing companies have a repository of VA resumes; organized and searchable by skill sets and competencies. In addition, virtual assistant staffing companies, like Team-Double Click, have internal training programs and rigorous screening processes of potential virtual assistants (VA's).

A virtual assistant staffing agency doesn't send you a "laundry list" of virtual assistants for you to "comb" through and find the right one. Virtual staffing agencies assess your personal needs as an agent and then match you with a trained and qualified virtual assistant. They take the process one step further by monitoring and nurturing the client/virtual assistant relationship so that you get the most from the assistant.

TeamDoubleClick.com

Team Double-Click exemplifies what it means to obtain a quality, virtual assistant and provides as a good case-study in this section. Besides the addition of transaction coordination and specialized transaction coordination training (TCTC) with the RELAY™ transaction management system, below is a comprehensive snapshot of what their independent contractors are commonly capable of performing (from the website of Team Double-Click): [23]

- Answer the telephone
- Forward calls
- Take messages
- Make appointments
- Build websites
- Put up and build PowerSites™ or single listing sites
- Send listing information to a multiple listing service
- Fill out necessary forms
- Deliver information and forms to a mortgage company and closing attorney or agent as part of the preparation for closing
- Make and deliver copies of public record
- Write and place advertising in newspaper and other forms of publication
- Receive and deposit funds to be held in trust for others including earnest money, security deposits, and rental payments
- Type forms
- Perform company bookkeeping
- Arrange for and oversee repairs
- Collect demographic information
- Solicit interest in engaging the services of a licensee or brokerage
- Set or confirm appointments (with no other discussion) for:
  o A licensee to list or show property

---

[23] *Professional Virtual Real Estate Assistants.* Team Double-Click, Inc. Retrieved December, 10 2005. http://www.teamdoubleclick.com/service_realestate.html. Reprinted with permission.

- o A buyer with a loan officer
- o A property inspector to inspect a home
- o A repair/maintenance person to perform repairs/maintenance
- o An appraiser to appraise property
- Make cold calls
- Make follow up calls
- Organize an open house
- Arrange closings
- Perform clerical duties
- Gather listing information
- Hand out preprinted, objective information
- Distribute information on listed properties when such information is prepared by a broker
- Deliver paperwork to other brokers
- Deliver paperwork to sellers or purchasers
- Deliver paperwork requiring signatures in regard to financing documents that are prepared by lending institutions
- Prepare market analyses for sellers or buyers on behalf of a broker
- Contact management
- Drip marketing campaigns
- Transaction coordination
- Personal tasks such as ordering groceries or making flight arrangements
- Many other duties too numerous to mention here [24]

Tips!

A VA staffing company, like Team Double-Click, Inc can provide screening, oversight of work-performed, and coordination of the payment process between the real estate professional and virtual assistant. Self-managed VA's do not have such oversight.

---

[24] *Professional Virtual Real Estate Assistants.* Team Double-Click, Inc. Retrieved December, 10 2005. http://www.teamdoubleclick.com/service_realestate.html. Reprinted with permission.

On the next two pages is an example of a Virtual Assistant's resumé. VA's are highly trained, educated individuals who look to play an administrative role to agents and brokers.

Sales Information

For sales information on obtaining a virtual assistant for the individual agent, broker, or to support agency tasks, visit www.TeamDoubleClick.com, call toll free at 888.827.9129 to speak with a sales representative, or contact by email at quotes@TeamDoubleClick.com.

NANCY P.

SUMMARY OF QUALIFICATIONS

- Over 10 years of business experience including sales, client services, operations and administration.
- Resourceful and detail oriented; skilled problem solver and multi-tasker
- Self-starter with excellent communication and business skills.
- TCRE certified by Team Double-Click, Inc.
- REAL ESTATE WEBOGRAPHER ™ professional[1]

PROFESSIONAL EXPERIENCE

September 2002-July 2005   Irvine, CA
Account Manager – LendingTree Loans/Home Loan Center Inc.
- Processed sub-prime alt-A and prime refinance and purchase loans.
- Maintained base of 40 to 50 clients.
- Effectively communicated with clients and ancillary departments to ensure smooth closing of loans.
- Provided excellent customer service to clients through attentive follow up.
- Consistently ranked within top 10% of processing staff.

March 2002-August 2002   Rosemead, CA
Account Executive – Full Spectrum Lending/Countrywide Home Loans
- Originated sub-prime, Alt-A and home equity loans by handling incoming phone calls and making outbound calls.
- Discussed and sold appropriate finance options to clients after evaluating their credit reports, financial status and property valuation.

December 2000-February 2002   Huntington Beach, CA
Human Resources Manager – E*Trade Mortgage
- Managed all aspects of Human Resources administration including developing and implementing policies/procedures; recruiting, interviewing and hiring personnel and conducting new hire orientation.
- Supervised staff of 2 Human Resources Assistants and a Benefits Administrator.

June 1998-December 2000   Huntington Beach, CA.
Loan Agent – E*Trade Mortgage
- Produced $6 million to $8 million in mortgage originations on a monthly

[1] National Institute of Webographers, LLC owns the certification marks REAL ESTATE WEBOGRAPHER ™, REW™ and REAL ESTATE WEBOGRAPHER (with W logo), which it awards to individuals who successfully complete initial and ongoing certification requirements.

Figure 2: Virtual Assistant Resumé—Sample *(page 1)*

basis; consistently ranked within top 10% of sales staff in production.

- Originated purchases, refinances, 2 nd trust deed products, A -minus and FHA transactions.
- Maintained a network of referrals by continuously corresponding with realtors as well as c urrent/previous clients.

June 1995-June 1998   Los Angeles, CA.
Recruiter/Administrative Specialist – UCLA Medical Center

- Recruited, interviewed and hired nurses and support staff.
- Completed employment verificat ions and background checks.
- Assisted in development of orientation and training programs for staff.
- Collected and analyzed data related to vacancy rates per unit, turnover statistics, projected hiring needs, referral sources and recruitment cost per hire.

EDUCATION

June 1995   University of California, Los Angeles   Los Angeles, CA.
Bachelor of Science in Psychobiology

OTHER SKILLS

- Licensed and Bonded Notary Public for the State of California
- Experience with RELAY Online Transaction Management – REBT.com
- Experience with ZipForm electronic forms
- Experience with Virtual Tours with RealBiz360 .com
- Experience with CMA/AVM generation with AgentAVM.com (eAppraiseIt)
- Experience with Agent/Agency websites – RapidListings.com
- Experience with Neighborhood Search inte gration – NeighborhoodScout .com
- Experience with Single Property Websites (123MapleSt.com) – AgencyLogic.com
- Experience with Microsoft Word, Excel & Power Point
- Experience with ACT!
- Experience in Mortgage Sales
- Experience in conducting Employment Verificati ons & Background Checks
- Experience working with Title/Escrow companies
- Experience working with Real Estate Attorneys across the country
- Skilled in internet research
- Experience in performing concierge services

Figure 3: Virtual Assistant Resumé—Sample *(page 2)*

PART TWO:
Your Initial Web Presence

Chapter 4: Webography

Chapter 5: Agent Website *(The Main Presence)*

Chapter
4

Webography

Webography is the art or practice of establishing a *seamless* web presence through the selection of web-based applications, joined together through basic and applied web techniques. Webography is an *evolution* from traditional web design. Websites and back-end functionality need not be created "from scratch" when following the Webography process. Why re-invent the wheel when hosting companies, service providers may provide an off-the-shelf application with a "point & click" interface? The model is appealing to self-serving web or Internet objectives without the hassle. Webography is a simpler process in creating and maintaining a web presence, and can be performed by the common computer user.

The Webography Process

Webography entails the use of pre-fabricated web pages and applications serviced by various technology providers that meet business objectives. These web pages for example, follow an initial theme like real estate, but require "slight" modification in appearance to create a brand image. Webography is a process. One first must define the requirements of the web presence. Essentially, what is the web site supposed to do? Secondly, what types of pre-built technologies and applications meet those requirements?

Thirdly, which service providers offer the required technologies and best integrate with one another for seamless functionality? Lastly, what are some techniques to integrate the look and feel of all selected components into one seamless web presence? These above questions are answered through a process called Webography. The Webography process consists of four (4) overarching phases as seen in Figure 4. Each phase consists of activities required to complete an entire web presence.

Figure 4: The Webography Process

Phase 1: Determine Your Requirements

In planning for an agent's web presence, it's important to establish clear-cut goals and requirements. What are you trying to achieve?

Overarching Questions

It's important to ask yourself initial planning questions during this process to establish your requirements. Questions about your intended web presence can include, but are not limited to the following:

1. Do I want a personal, real estate web presence to showcase my own properties as the listing agent?
2. Do I want to include other MLS listed properties from my local area in my web presence?
3. Do I want my web presence to generate leads?
4. Do I want to provide access to transactional documents from my web presence?
5. Do I want to provide tools that are appealing to out-of-town buyers?
6. Do I want "ease" in maintaining the solutions myself?
7. Do I want to access parts of my web presence from a mobile device?
8. Will I have time to manage all this myself or should I hire an assistant?

Figure 5: Webography—Phase 1

Questions into Requirements

Given the overviews of technology services stated in Chapter 1, and specific details in subsequent chapters, you have an overarching idea of web-based technologies that can meet your business requirements. Let's play the role of an agent asking these questions of him or herself:

1. *Do I want a personal, real estate web presence to showcase my own properties as the listing agent?* Yes, I'd like to better engage potential buyers with my own web presence and build a professional image around myself and my personal skills. Although an agent within a brokerage, I myself need to establish a "business" around my name.

2. *Do I want to include other MLS listed properties from my local area in my web presence?* Yes, I'd like to also incorporate local properties of other agent's to further appeal to potential buyers perusing my web presence.

3. *Do I want my web presence to generate leads?* Yes, I want the traffic of site visitors to provide me contacts and leads I can follow-up with.

4. *Do I want to provide access to transactional documents from my web presence?* Yes, I do want to better organize my transactional documents for myself and for my clients using some type of online application or interface.

5. *Do I want to provide tools that are appealing to out-of-town buyers?* Yes, and I do understand that web-based technology can bring a sense of "realism" to the properties I display within my web presence. I do understand my listing process must embrace the Internet and web-based technologies.

6. *Do I want "ease" in maintaining the solution myself?* Yes, and I do know many of today's web-based applications are built where I don't need to be a computer programmer. I've heard of many web-based applications having "point & click" configuration, where I don't need a traditional web designer, nor be one myself.

7. *Do I want to access parts of my web presence from a mobile device?* Yes, I know many of real estate technologies allow for editing or viewing capabilities from a mobile device like a PDA.

8. *Will I have time to manage all this myself or should I hire an assistant?* Unsure, as last year my sales were average and would like to double sales this year. However, all this technology could pull me away from the focal task of getting new clients and making sales? Maybe I should look into finding an assistant who could help maintain my technology.

Requirements Have Been Established
From this exercise, this agent has concluded "yes" to all brainstorming questions to begin to define a path towards a finalized web presence.

Phase 2: Select Required Technologies

The next step in the Webography process prompts one to translate the requirements into supporting technologies. Knowledge of today's popular real estate technologies comes from reading the remainder of this book, in addition to completing the REAL ESTATE WEBOGRAPHER™ certification at www.Webographers.com. Given that knowledge, one can apply a technology-type with each requirement.

Figure 6: Webography—Phase 2

1. *Do I want a personal, real estate web presence to list my own properties as the listing agent? Yes.*
   a. Agent Website
      i. Domain name
      ii. Email addresses
   b. Single Property Websites
      i. Domain names
         (for each property)
   c. Online Classifieds for Listings
      (Local newspaper online, Realtor.com, Yahoo! Real Estate, etc)
2. *Do I want to include other MLS listed properties from my local area in my web presence? Yes.*
   a. Agent Website

    i. MLS IDX capability *or*
    ii. MLS raw data feed

3. *Do I want my web presence to generate leads? Yes.*
  a. Agent Website
    i. Contact/Lead web forms
    ii. "Refer a Friend" functionality
    iii. Business Contact Information
    iv. Ability to hyperlink or "frame" 3rd party web pages such as:
      1. Neighborhood Search
      2. CMA (AVM) report obtained instantaneously by site visitors.
  b. Single-property websites
    i. Contact/Lead Forms
  c. Online Classifieds
    i. Contact/Lead Forms
  d. Virtual Tours
    i. Contact/Lead Forms
    ii. "Refer a Friend" functionality
    iii. Personal Contact Information

4. *Do I want to provide access to transactional documents from my web presence?*
  a. Online Transaction Management (OTM)
  b. Electronic Forms Provider
    i. "Online version" for mobility in accessing/editing forms
    ii. Can be Integrated with my OTM Provider
  c. Agent Website
    i. Ability to "link" to the OTM "landing" page from my agent website.
  d. Single-property Websites
    i. Ability to "link" to the OTM "landing" page from each Single-property website.

5. *Do I want to provide tools that are appealing to out-of-town buyers? Yes.*
  a. Agent Website
    i. Property Listings with Photos
    ii. Virtual Tours—24x7 open house
    iii. Neighborhood Search for Communities
    iv. CMA (AVM) report for property valuation and comparable sales.
  b. Single Property Websites
    i. Property Listings with Photos

         ii. Virtual Tours—24x7 open house
   c. Online Classifieds
         i. Property Listings with Photos, URLs
   d. Virtual Tours—24x7 open house
         i. Ability to post tour on Realtor.com and other online "classi-fied outlets".

6. *Do I want "ease" in maintaining the solution myself? Yes.*
   a. All technologies should be "point & click", with no HTML knowledge required.
   b. Should also look for technologies "natively" integrated with each other; having partnerships.

7. *Do I want to access my web presence from a mobile device? Yes.*
   a. Agent Website
         i. Receive email leads
   b. Mobile Application (ideally allowing for the following)
         i. MLS property listing information
         ii. Virtual Tours
         iii. Documents
         iv. Location Maps
         v. Real Estate Calculators

8. *Will I have time to manage all this myself or should I hire an assistant? Unsure.*
   a. Virtual Assistant staffing company who trains many of their VA's in similar technologies that I may be acquiring.

Technologies Have Been Selected

For this agent, a given *technology* may support many *requirements*. To better organize the selected technologies, this agent lists each technology once with the desired features. After organizing the selected technologies this agent has decided upon the following listed "a" through "j":

a. Agent Website
- Property Listings with Photos
- MLS IDX capability or MLS raw data feed
- Contact/Lead forms
- "Point & Click" interface (no programming)
- Ability to hyperlink or frame in 3rd party web pages.

b. Online Transaction Management—subscription

c. Electronic Forms (online version) provider
   - Integrated with OTM

d. Neighborhood Search—subscription

e. CMA (AVM) Report—subscription

f. Virtual Tour—subscription
   - Post to any website
   - PicturePath™ provider to post tours on Realtor.com

g. Single-property Websites—package of many websites

h. Mobile technology application

i. Virtual Assistant company—locate potential provider

j. Online classified outlets (as needed)

Phase 3: Select Service Providers

The next step in the Webography process is to find service providers that best fulfill the technology needs established in Phase 2. This 3rd phase may be a challenge if you are unfamiliar with service providers of real estate technologies and applications. Subsequent chapters in this book delve into great detail on real estate web applications and technologies found in today's market. Technologies highlighted by the Corporate Sponsors of the REAL ESTATE WEBOGRAPHER™ certification, are showcased to provide "real" and commercially available solutions.

Is there one service provider that can provide all the selected technologies from Phase 2? Regretfully, there is no "one solution fits all". When this is the case, you should first find a service provider that can provide the majority of your selected technologies from Phase 2. This service provider establishes what is known as the Main Presence. As shown in Figure 7 on the next page, "Service Provider A" provides most of the technology features desired by the agent.

This Main Presence is the "storefront" to an agent's web presence and is commonly accessed through the agent's personal URL (i.e. www.<TheAgentsWebsite>.com) The remaining technologies not covered in the Main Presence will be supplied by additional service providers that best integrate with the Main Presence.

Figure 7: Webography—Phase 3

Agent's Main Presence

As noted during Phase 2: Select Required Technologies, the *agent website* encompassed many of the desired features, and was the "entryway" for visitors to enter you web presence and access all inclusive services. The agent website should be viewed as your <u>Main Presence</u>, so selection of an agent website provider is critical and should be your first step.

1. Agent Website (*Main Presence*)
    a. Domain name and email
    b. Property Listings with Photos
    c. MLS IDX capability *or*
    d. MLS raw data feed
    e. Contact/Lead forms
    f. Ability to hyperlink or frame in 3rd party web pages such as:
        i.   Online transaction management
        ii. Neighborhood Search
        iii. CMA (AVM) report for property valuation and comparable sales.
        iv. MLS IDX search page

    Provider Selected—**RapidListings.com**

Other Technologies

An agent must establish customer relationships with various service providers, being mindful of how they can be joined together for a seamless functionality, look, and appearance. Do they play well with each other? When it is time to select service providers of other required technologies, it's important to be mindful of later establishing a seamless web presence:

- Selecting products natively compatible with each other
- Ensuring the same personal and corporate branding can be applied across all platforms
- Selection of similar color schemes is available, etc

2. Electronic Forms (online version) provider
   Provider Selected—**ZipForm***Online*

3. Online Transaction Mgmt subscription
   Provider Selected—**RELAY™**

Note. With the selection of ZipForm*Online* and RELAY, it was noted both applications are natively integrated and compatible with each other. Forms can be completed at ZipForm*Online*, and then seamlessly "exported" to a RELAY, transaction management, agent account.

4. Neighborhood Search subscription
   Provider Selected—**NeighborhoodScout®**

Note. A NeighborhoodScout, search page can be added to any RapidListings' agent website, by linking or framing the web page, as noted by this agent.

5. CMA (AVM) report subscription
   Provider Selected—**AgentAvm.com**

Note. AgentAvm.com access page for site visitors can be added to any RapidListings' agent website by linking or framing the web page.

6. Single-property websites package
   a. Domain name and email
   b. Property Listings with Photos
   c. Domain names for every property website.
   Provider Selected—**AgencyLogic.com**

Note. AgencyLogic provides for the creation and publishing of single property websites, and inclusive services to obtain a domain name (i.e. 123AnySt.com) for every listing. In addition, each AgencyLogic PowerSite URL (i.e. http://www.123AnySt.com) can be cross-referenced on their RapidListings.com agent website.

7. Virtual Tour—subscription

    a.   Post to any website

    b.   PicturePath™ provider to post tours on Realtor.com
    Provider Selected—**RealBiz360.com**

Note. RealBiz360 can allow one to "publish" a virtual tour to another website (i.e their RapidListings agent website and Agengylogic Single-Property website. In addition, as a PicturePath™ provider, virtual tours can be published directly to Realtor.com.

8. Mobile Technology Provider

    a.   Property Listing Information

    b.   Virtual Tours

    c.   Documents

    d.   Location Maps

    e.   Real Estate Calculators
    Provider Selected—**iseemedia's iseerealty**

Note. This agent noted that iseerealty provides access to Property Listing Information, Virtual Tours, Documents, Location Maps, and Real Estate Calculators on mobile phones and wireless PDAs. Coming soon to RealBiz360.com, iseerealty (developed by iseemedia) is seamlessly integrated into RealBiz360.com's essential Virtual Tour, marketing application.

9. Virtual Assistant company—locate potential provider
    Provider Selected—**Team Double-Click, Inc.**

Note. This agent has noted that Team Double-Click train their VA's on many of products & services of the providers mentioned above.

10. Online classified ad postings (as needed)
    Provider Selected—Realtor.com, Yahoo! *Real Estate*, local newspaper.

Note. Due to the efforts of preparing images and virtual tours for the agent website and single property websites, this agent notes that the work is not wasted and can be reused in real estate classified outlets.

<u>Service Providers Have Been Selected</u>
All service providers and their respective technologies have been selected. These services will encompass the all-inclusive web presence of the agent.

Phase 4: Establish Seamless Appearance

The final step in the Webography process is to piece all the technologies together into one seamless web presence.

This phase requires creating a look that is seamless to the end-user. This can be achieved by ensuring company information and brand (i.e contact information, company logo) appear on any of the services encompassed in the web presence. When applicable, adding your agent photo, agent contact information, and biographical information is imperative. In addition, color schemes and any artistic work should be in-concert throughout all services a user may come across. These techniques should provide as the "glue" that binds together all technologies of the service providers.

Figure 8: Webography—Phase 4

The milestones in the web design process mentioned above have significantly reduced the man-hours and know-how to develop functional web sites and pages. Overall, the 4 phases of Webography provide a step-by-step process in establishing a web presence. Technology today provides for the ability to create a web presence with little web design skills or none at all. The next chapters detail many of the technologies that can be inclusive of an agent's web presence.

Webography Certifications

As defined by the National Institute of Webographers, LLC (NIW), Webography is the art or practice of establishing a *seamless* web presence through the selection of web-based applications, joined together through basic and applied web techniques. The NIW focuses on the practice of Webography, currently for the market of real estate with its flagship certification called REAL ESTATE WEBOGRAPHER™ certification. The service of training, assessment, and certification in Webography for real estate professionals is provided exclusively by the National Institute of Webographers, LLC at Webographers.com.

Agent Website

(The Main Presence)

For real estate agents, establishing a web presence includes selecting a service provider that fulfills most, if not all, technology requirements. This is later followed by selecting service providers that fulfill any remaining technology needs that encompass your entire web presence. The Main Presence is represented as *www.YourAgentWebsite.tld*.

Discussed are traditional hosting companies that provide tools to host a website, a website developed from scratch. This "history" is presented to show why Webography has evolved into what it is today. A *Webographer-friendly hosting company* provides pre-built web pages and inclusive tools to get a website up and running in a very short amount of time. This chapter details that evolution.

Before we establish a Main Presence to support real estate objectives, let's discuss how you'll access the Internet. Selecting a service provider for Internet access is the first step in determining how to best utilize Internet applications for real estate objectives.

Internet Service Provider

One will conclude from reading this book that activities performed by a REAL ESTATE WEBOGRAPHER™ can be Internet-intensive. It's important to look at various types of Internet connections, as connection speeds may propel or adversely inhibit many Internet activities. Establishing and maintaining a web presence and inclusive applications should not be deterred by slow Internet connection speeds.

Those who desire to use a standard 56k modem will experience frustrations in performing web-based tasks and objectives. High Speed (or Broadband) is the recommended group of Internet connection speeds that should be utilized by a REAL ESTATE WEBOGRAPHER™.

Internet connection speeds are generalized in the chart below:

ISP Connection Speeds (estimate)							
Data Stream	28.8K Modem	56K Modem	56K-64K ISDN	112K - 128K ISDN	256K Partial T1 or DSL	512K Partial T1 or DSL	1.54 Mb T1
1 Mb	5 min	4 min	2 min	1 min	1 min	20 sec	6 sec
5 Mb	25 min	21 min	11 min	6 min	3 min	1.5 min	30 sec
10 Mb	49 min	42 min	22 min	11 min	6 min	3 min	54 sec
20 Mb	1 hr 40 mins	1 hr 25 min	45 min	20 min	11 min	6 min	2 min
50 Mb	4 hr	3 hr 45 min	1 hr 50 min	55 min	27 min	14 min	4.5 min
100 Mb	8 hr	7 hr 30 min	3 hr 40 min	1 hr 50 min	55 min	27 min	9 min

Dial-up. Connections that require the modem to "dial" a phone number for each connected Internet session falls into the 28.8K to 56K modem connections. This entails establishing a connection with the service provider through the sequence of a modem dialing a phone number. This process adds to the time involved in performing functions with online applications. This connection is not always live and can be "cut" by an inbound call.

Broadband. Typically, broadband (high speed) connections fall into the 128K to 1.54 MB connection as shown above. Such connections are always "on", not requiring the manual process of dialing and disconnecting connections at every sitting. These types of connections are highly recommended for the REAL ESTATE WEBOGRAPHER™ professional.

In the recent 2005 REALTOR® Technology Efficiency Study, a survey completed by 2554 of NAR members, 87% of the respondents mentioned using broadband or high-speed connections. 12% mentioned using dial-up[25]. Thus the numbers of real estate professionals using high-speed Internet access, lends itself to more use of extensive Internet-based applications.

When selecting an Internet Service Provider (ISP) to provide broandband (high speed) Internet access, one must consider these 6 items:

1. *Purpose.* Are you simply using the Internet for email purposes, or for more extensive activities like managing property listings on your agent website (upload a 0.5 MB image) or online transaction management? Uploading a 0.5 MB photo from a digital camera to your agent website would take close to 2 minutes, where a 512k DSL connection would take 10 seconds!

2. *Support.* Does the ISP (Internet Service Provider) offer 24x7 customer support? Do they offer set-up support assistance over the phone? Do they offer "tiered" technical support for resolution to active ISP subscriptions?

3. *Connection.* What are the various connection speeds offered? For example, an ISP saying they provide DSL is not specific enough. Is their stratification within the DSL service, i.e. 256K or 512K?

4. *Reliability.* What guarantees does the ISP make to uptime of their service? What do customers say about this ISP? Are their testimonials or reviews in online publications about the reliability of this ISP?

5. *Price.* How does the pricing of this ISP compare to other providers that service your local area? Do they offer bundled packages that include high-speed Internet service and business or home phone at a discounted rate?

6. *Business.* Does the ISP also have partnerships with local businesses that provide wireless "hotspots"? In consideration of mobile technologies, can you use your ISP account (username & password) at your local bookstore to access the Internet at no additional charge? A very affective resource when meeting clients at "middle ground" locations.

Let's suppose you have established an effective platform to access the Internet for web-based activities. Hopefully, you have established high-speed (broadband) access with an ISP. In establishing the initial web presence (i.e. GregDugan.com), let's now examine providers of agent websites.

---

[25] 2005 REALTOR Technology Efficiency Study. (2005, May 03), pgs 5, 10. http://www.Realtor.org/CRT

Traditional Hosting Companies

It's important to first examine hosting companies that provide general solutions to establish an agent website. These traditional hosting companies are not recommended, unless traditional web design and programming are to be performed by a designer you have hired. This type of provider is discussed to show the "history" of web design and its evolution. This section examines the movement from hosting services appealing to *web designers* to services that appeal to the *webographers*.

There are hosting companies that provide solutions for typical web designers who build custom websites from scratch. Their platforms are historically for developing a real estate website from the ground-up, not typically providing many of the inclusive functionality or pre-built web pages that a *Webographer-friendly hosting company* would provide like RapidListings.com, the case-study service provider showcased in the remainder of this chapter.

Traditional hosting companies include customer platforms that are not designed to cater to specific markets, but more so to the masses. Thus their platform must be general, not specific to real estate for example. Mainly, traditional hosting companies appeal to those who desire to perform custom website development.

Traditionally speaking, such companies present their services in terms of disk space, bandwidth transfer, types of databases supported, etc. However, those who want to create more custom websites in support of real estate web objectives should examine the "Traditional Hosting Companies" overview below.

A REAL ESTATE WEBOGRAPHER™ professional following the typical Webography process as mentioned in Chapter 4, is not really worried about these features as shown below. As one can see above, those who desire to establish an effective web presence may not be familiar with those features of traditional hosting companies. This is because traditional web hosting companies cater to web designers and <u>not</u> webographers. The next section examines *Webographer-friendly hosting companies*, those that cater to webographers.

Traditional Hosting Companies	
Disk Space & Traffic	Allowed Disk Space, bandwidth traffic, hits per month.
Domain Name	Selection of a desired domain names and ability to establish sub-domains.
Site Management Tools	24/7 FTP access, Web-based control panel, Microsoft® FrontPage® extensions 2002, Detailed traffic stats (Analog), Access to raw log file
Email	POP3 / SMTP accounts, Web-based email, Spam and virus blocker, email forwarding, email aliases, autoresponders, inclusive mailing list tools
Platform Type	Linux/Unix or Windows Platform
Free Software / Scripts	Blogs, Counters, FormMail, Guestbooks, Message Boards
Databases	PHP, MySQL, PHPAdmin
Multimedia	Shockwave, Flash
Security Features	SSL (Secure Socket Layer) encryption, SSH (Secure Shell / Telnet) encryption
Support and Backup	24/7 support via toll-free phone & email, Daily data backup

Webographer-friendly Hosting Companies

Hosting companies that are Webography-friendly present their services in terms of themes, like real estate, with pre-built web services, pages and applications that support that theme. They provide established solutions for the real estate market, where a customer selects a desired template, and then modifies inclusive, pre-fabricated web pages, etc. to meet their needs and objectives. Instead of discussing disk space, bandwidth restrictions, etc, *Webographer-friendly hosting companies* speak of website packages that are specific to a market like real estate. Such features may include: number of property listings per website, number of images per property listing, available pre-built web pages like "Buyer Resources" or "Seller Resources", etc.

Figure 9: Agent Website—Powered by RapidListings.com[26]

Next, we'll continue discussion on how one utilizes a *Webographer-friendly hosting company*, and maintains a web-presence through inclusive administration tools.

Control Panel

A *Webographer-friendly hosting company* will provide a "Control Panel" that allows you to manage a website with no HTML or coding required. In the control panel, an agent selects what pre-fabricated web pages they'd like to include, edit web page content with the *In-browser, web page editor*, upload images, send mass emails, etc. This "point & click" interface allows agents to instantaneously modify, update, and change the website from any computer with Internet access.

Templates

Commonly, the control panel is where you select the template (look) of the website. These templates are designed with a real estate theme, where the agent can select a desired "look & feel" that they later make their own. In comparison to traditional hosting companies, inclusive functionality is already configured for you, allowing the REAL ESTATE WEBOGRAPER™ to have a website ready in minutes. With such ease in set-up and maintenance, an agent can self-serve their own agent website. Figure 10 on the next page showcases a sub-set of templates from such a hosting company that agent users can select.

---

[26] *Client Showcase.* RapidListings.com. Retrieved December, 10 2005. Reprinted with permission. http://www.RapidListings.com

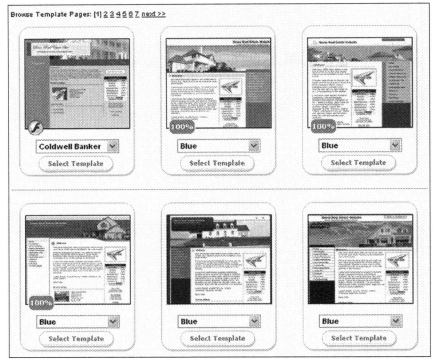

Figure 10: Templates—RapidListings.com Agent Website [27]

In-Browser, Web Page Editor

As discussed above, the Control Panel commonly includes an editor to edit & save web pages, with changes applied immediately to the website. An *In-browser, web page editor*, is an editor that looks like a word processor and is accessed from a web page. Commonly, one can add text and perform functions on text like bold, underline, and other common word processor functions. Essentially, the website can be edited from any computer with Internet access. Updates and changes are immediately saved, with no transfer of HTML files.

An *In-browser, web page editor* may be deemed as requirement for simplicity in setup, design and future maintenance. This is true for pages that are static, mainly

---

[27] *Change Template.* RapidListings.com. Retrieved December, 10 2005. Reprinted with permission. http://www.agentedit.com/edit_website_content.asp.

text-based. Such pages may be "Home Page", "News", "Specials", etc. An *In-browser, web page editor* provides all the required features as if you had a web development application on your desktop. The interface to edit web pages is browser-based, and accessible from any PC with an Internet connection.

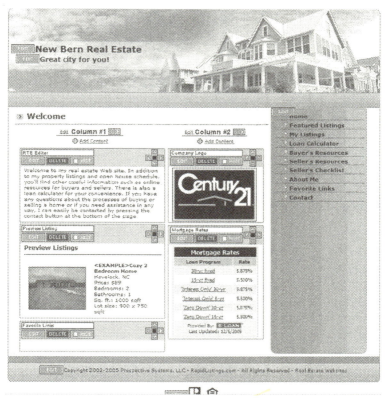

Figure 11: In-browser, Web Page Editor—RapidListings.com Agent Website [28]

From an independent contractor perspective, those who establish but may not maintain a web presence for agents, the *In-browser, web page editor* empowers clients to self-maintain the web site after initial layout is established. For agents

---

[28] *Edit Website Content*. RapidListings.com. Retrieved December, 10 2005. Reprinted with permission. http://www.agentedit.com/edit_front_page.asp.

who look to use Virtual Assistants, the *In-browser, web page editor* enables those assistants to make updates at anytime from anywhere.

When a user of such an editor clicks "Save" in the web page editor, the page is automatically saved to the website instantaneously. It's this simplicity that inspires self-serving maintenance by real estate professionals. An *In-browser, web page editor* should have the following capabilities:

- Image Control—ability to modify Alt Tag, horizontal and vertical spacing, scale.

- Table Control—easy modification of table & cell properties: cell height, rowspan, colspan, visual border and background settings and No-text-wrap features.

- Web page Control—edit current links, Alt tags and targets, set absolute vs. relative paths, format text like a word processor, and password protect the editor for added security.

Inclusive Lead Generation

Company websites, followed closely by <u>agent</u> and MLS public websites are the largest sources of Internet leads.[29] Native to agent websites provided by *Webographer-friendly hosting companies* are resources for agents to acquire leads from site visitors. Figure 12 on the next page describes the lead generation tools included with an agent's website hosted with RapidListings.com. In terms of a *Webographer-friendly hosting companies*, it's assumed that features like Contact Forms, "Refer a Friend," inquiry forms for "Home Value Analysis" are included. None of these capabilities require additional programming, as their included and pre-configured. A RapidListings.com, agent website can even email leads to your cell phone's associated email address.

---

[29] 2005 REALTOR Technology Efficiency Study. (03 May 2005), WAV Group in conjunction with Center for REALTOR® Technology (CRT). p14. http://www.Realtor.org/CRT

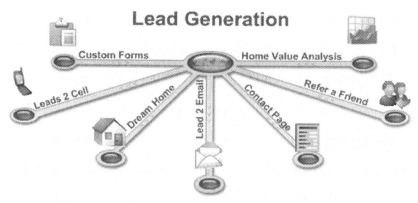

Figure 12: Lead Generation Capabilities at RapidListings.com[30]

Activity!

At Webographers.com, each candidate of the REAL ESTATE WEBOGRAPHER™ certification is supplied their own RapidListings.com, agent website. Here, candidates can practice working with adding property listings, working with lead generation tools, and receive leads from a potential seller (course proctor) to simulate electronic communication mitigated from an agent website.

Domain Name

The domain name is the entryway into an agent's web presence. The URL typed into the user's browser, begins the start of a unique and agent personalized, online experience. Selecting a domain name is important and should not be rushed. It's the gateway to your web presence. Domain name selection and acquisition should not be taken lightly.

Availability

You must understand that your desired domain name may not be available. A REAL ESTATE WEBOGRAPHER™ professional is capable of brainstorming

---

[30] *Lead Generation.* RapidListings.com. Retrieved December, 10 2005. Reprinted with permission. http://www.rapidlistings.com/features/lg/

other names, following proper Domain Name "etiquette". Please consult a *Webographer-friendly hosting company* (like RapidListings.com) for information on obtaining a domain name to use with your agent website.

Domain Name Etiquette
Selection of domain names should try to refrain from the use of hyphens. In terms of businesses, .com's or always preferred, although .net, .us, .ca (Canada) should be considered if the desired name is unavailable. The shortest number of characters in the domain name is always preferred.

Consider the "Doing Business As" (DBA) name of many small businesses. Normally, some part of the DBA name appears in the domain name as to not confuse customers, and make the name easy to remember. For example, the DBA name of the copyright holder of this book is the *National Institute of Webographers, LLC*. The domain name of Webographers.com, is short in length and uses some part of the registrants DBA name.

How does this method apply to real estate agents and their domain name? Given the goal of establishing a web presence around the agent's good name, you may consider your full name like that of a business's DBA name. Using some or your entire birth name in the domain name also helps users easily remember the domain. If this method is unfruitful in finding an available domain name, the domain name should imply real estate and possibly the geographic location where you're located.

Tips!

URLs entered into an Internet Browser are <u>not</u> case-sensitive. It's best to use capital letters for the first letter of each word appearing in your domain name on all marketing and advertising material when displaying a domain name.

Displaying a Domain Name
Let's consider that the selected domain name has many inclusive phrases. Consider that the REAL ESTATE WEBOGRAPHER™ certification had its own domain name. How does www.realestatewebographer.com read to you at first glance?

You should be saying it doesn't read well! It's long and at first glance, hard to distinguish the inclusive phrases, making the domain name (URL) hard to remember. It should be displayed as follows: www.RealEstateWebographer.com. Here, capitalizing the first letter of each inclusive word or phrase improves readability. In addition, it helps users better remember your domain name.

- Bad Display
  www.agencyabc.com
  www.gregdugan.com
  www.realestatewebographer.com

- Good Display
  www.AgencyABC.com
  www.GregDugan.com
  www.RealEstateWebographer.com

If space is limited, like on a business card, feel free to drop the "www" when displaying the domain name (i.e. GregDugan.com). When users enter the URL in a browser, not placing www in the navigation bar of the web-browser should still resolve to your agent website.

Signage

It's known that signage placed near the curb-side of listed properties appeals to passer-bys browsing a neighborhood. The content displayed on the sign not only announces the property is for-sale, but can also effectively advertise the agent. For those considering selling their home, they may see an agent's signage throughout a local area. Such advertisement of the agent readily viewed by passer-bys implies the agent is a desired commodity, locally renowned for their services.

Figure 13: Sign with URL of Agent Website

An agent website is only as effective as the number of people who know about it. Although signage may reflect the agency's branding, it's also a perfect place to make known the domain name of an agent website. Given Figure 13 on the previous page, the agent effectively mention's his name "Greg Dugan", and his website, in one (1) powerful statement.

Business Email

Besides the basic understanding that email consists of *username@domain_name.tld* for example, it's important for REAL ESTATE WEBOGRAPHER™ professionals to understand email in terms of client impressions. A business email is an extension of a selected domain name. It's a very exciting to achieve a desired domain name i.e. GregDugan.com for an agent's personal web presence. An email using that domain name is an extension of that excitement, Greg@GregDugan.com.

This is true for real estate professionals who want to display a professional image using a business email. Hosting companies should always provide a web-based interface to check email, also to coincide with on-the-go, mobile technologies. RapidListings provides inclusive email accounts with every agent website if so desired.

Client Expectations

Many real estate professionals may use free email addresses like Hotmail, Yahoo! or through their Internet Service Provider (ISP). It's very hard to simply find an available username that includes 'first name' and 'last name' with such providers (i.e. JohnDoe@hotmail.com). Using free email, or one through your Internet Service Provider, also detracts from your business edge while reducing your brand image.

> **Caution!**
>
> If you're a real estate professional using a free email address, or an address through your ISP for business communication, you should strive to quickly change to a business email.

Client Trust

What sounds more professional to you in the long run, JohnDoe121@hotmail.com or JohnDoe@AgencyABC.com? Who's email sent to you would you trust more, from JohnDoe121@hotmail.com or JohnDoe@AgencyABC.com?

Activity!

Candidates of the REAL ESTATE WEBOGRAPHER™ certification at Webographers.com, are supplied a basic email account (JDoe@Webographers.com) to engage in best practice training with clients (course proctors). In addition, candidates learn to navigate through automated email messages generated by the technologies offered in-part by Corporate Sponsors.

Again, the email address is an extension of your agent website. In terms of referral business, if a customer is forwarded an email from Greg5564@hotmail.com (Greg Dugan, a real estate agent), there is nothing that redirects them to his web presence. However, a person who receives an email from Greg@GregDugan.com. knows immediately that an agent website really exists, because "GregDugan.com" is displayed in the email address!

Response Time to Email
The Center for REALTOR® Technology states that many agents do not understand that email leads/questions need fast and efficient responses. They express that agents likely have 12 hours to respond to email communication from potential clients before they become distraught, possibly even giving a lead to an agent who responds more timely.[31]

Maintaining Relationships with Past Clients
Chances are, you're using some form of CRM (Contact Relationship Management) such as Microsoft Outlook, Act!, Top Producer, or even web-based email with an inclusive address book to help manage your vast list of contacts. Besides CRM use for sending and organizing emails, CRM tools help to organize past clients' contact information; those who could provide repeat business. Reports by NAR also state the top three methods to stay in touch with past clients include: Mailings (40%), Phone calls (31%), and Email (17%). About 65% of the respondents say referrals and repeat business account for more than 25% of their business.[32]

---

[31] 2005 REALTOR Technology Efficiency Study. (23 Feb 2005), WAV Group in conjunction with Center for REALTOR® Technology (CRT). p30. http://www.Realtor.org/CRT
[32] 2005 REALTOR Technology Efficiency Study. (2004, May 03), WAV Group in conjunction with Center for REALTOR® Technology (CRT). p6. http://www.Realtor.org/CRT

As one can see, selection of the initial steps taken to establish the initial web presence, with inclusive email is extremely important. Such care taken during these initial steps set the stage for an effective web presence. There are few providers who can enable all the features and ease-of-use for agents to establish the Main Presence. The next few chapters detail technologies to further extend the web presence of an agent.

RapidListings.com

Is the one of the nation's leaders in websites for agents and brokers. Through an easy-to-use control panel, websites can be tailored to meet the needs and personal interests of each agent. By deeming RapidListings as a *Webographer-friendly hosting company*, agents and brokers can have a site up and running with little-to-no HTML knowledge.

Besides agent websites, RapidListings provides resources to obtain a domain name for your website, and supply you with email accounts.

Sales Information

For sales information on establishing agent or broker website at RapidListings account, visit www.RapidListings.com, call toll free at 1-800-798-1838 opt. 2 to speak with a sales representative, or by email at sales@Rapidlistings.com. A comprehensive overview of various agent/broker website packages found at RapidListings.com includes:[33]

RapidListings—Website Features	
Custom Pages	Create custom pages using an in-browser web site development editor
Pictures Limited or un-limited	Upload as specified number of pictures per listing, or have an unlimted number of pictures
Domain Name	A domain for the agent / agency website, i.e AgencyABC.com
Email Accounts	The number of inclusive email accounts, i.e. John@AgencyABC.com
Contact Group Size	The number of recipients that can be included in an email group.
Editable Listing Brochure	A brochure of a property listing on the website can be generated as a PDF, or generated in an "editable" format.
Changeable Templates	The ability to change the look of a website at any time by selecting a desired template
Featured Listings	A resource to select properties to highlight on the main page of the website.
Unlimited Text	Ability to provide the required amount of text on a given page of the website.
Turn Pages off/on	Allows you to "turn off" a web page instead of deleting a page. Interacts with navigational buttons, removes a button when a web page is turned off.
Auto Mapping Feature	Links address of property listing to a mapping website, i.e. Google Maps to generate a map of the property's location.
Instant Changes	No transfer of web page files required, changes to site are immediately updated.
Rich Text Editor	Ability to create bold, underlined, italic text. In addition to text of various colors.
Auto Image Resizer	Ability to not have to resize or shape an image before uploading to the website, system does that for you.

---

[33] *Pricing/Features.* RapidListings.com. Retrieved December, 10 2005. Reprinted with permission. http://www.rapidlistings.com/pricing.asp.

Mortgage Calculator	A Value-Added, pre-designed web page that provides users the ability to compute mortgage costs.
Links Page	A Value-Added, pre-designed web page that provides for input of links to relevant websites
Buyer's Resources	A Value-Added, pre-designed web page that provides knowledge and tips to potential buyers.
Seller's Resources	A Value-Added, pre-designed web page that provides knowledge and tips to potential buyers.
Contact Page	A Value-Added, pre-designed web page that provides all forms of contact of the agent or agency, i.e. office address, phone numbers, fax numbers, etc.
Contact Generation Forms	A Value-Added, pre-designed web page that provides potential customers to enter in contact information. Commonly the user data is automatically entered into the contact management resource of an agent.
Search Engine Friendly	Tools provided to optimize the website to achieve the best possible search engine rankings.
Seller's Checklist	A Value-Added, pre-designed web page that provides a checklist of "to-do" items to potential sellers.
Virtual Tour Compatible	Ability to hyperlink Virtual Tours hosted with another provider directly with property listings displayed on the website.
Meta-Tag Controls	Ability to add a title and keywords that describe your website. These phrases play a part in search engine results.
Real Estate FAQ	A Value-Added, pre-designed web page that provides knowledge to questions commonly asked of customers.
Real Estate Glossary	A Value-Added, pre-designed web page that provides definitions of real estate terminology that customers may not be familiar with.
Buying Mistakes	A Value-Added, pre-designed web page that describes activities that buyers may want to avoid.
Selling Mistakes	A Value-Added, pre-designed web page that describes activities that sellerss may want to avoid.
Testimonial Page	A Value-Added, pre-designed web page where agents or the agency can list positive statements from previous customers.
Real Estate News	Integrated news feed from top real estate news publishers.
Search Listings	Integrated search engine that provides for users to search agent / agency listed properties based on fields (i.e. number of bedrooms, number of bathrooms, etc)

Flash Templates	Ability to integrate animated features to the website.
Navigational Re-arrangement	Ability to rearrange displayed buttons that hyperlinks to various pages of the website.
Custom Templates	Service-Provider may provide inclusive web design services to modify their existing web site templates, or create a new one.
Separate Rental Page	A Value-Added, pre-designed web page that lists properties that are "For-Rent" only.
Separate Storage Rental Page	A Value-Added, pre-designed web page that lists storage facilities associated or in partnership with the company.
Listing Share Program	Ability to cross-list property listings with other agent / or agency websites that are hosted with the service provider.
Re-Name Navigation Buttons	Ability to rename the titles that appear on buttons to pre-designed web pages.
Custom Forms	Service-provider may provide additional services of designing custom forms for various purposes.
IDX/MLS Integration	Ability to diplay listings form the local MLS systems. Usually required a hyperlink (URL) from the local MLS.
Dream Home Finder	A Value-Added, pre-designed web page includes a form for users to enter in exactly what they are looking for in a home. Contact information is then added to a contact manager resource.
Multiple Agent Profile Page	A Value Added resource for agents/brokers that includes many web pages that provides profiles of inclusive agents.
Listing Agent Designation	For each listing, an agent photo, name, contact information, and possibly company logo are provided with each property listing.
Re-Organize Agents	Ability to edit the order agents are listed on the "Agents" web page.
Setup Fee	The fee that is paid to establish an account with the provider.
Price (Monthly Fee)	Fee that is paid per month to have the site hosted.

PART THREE:
Connect Consumers through Internet Marketing

Chapter 6: Bring Traffic to an Agent Website

Chapter 7: Single Property Websites

Chapter 8: Virtual Tours

Chapter 6

Bring Traffic to an Agent Website

As discussed in the previous chapter, there are many techniques to bring traffic to an agent's website, especially those you come into *human* contact. Other techniques look at search engines and other websites that may reference your website.

The Human Touch
Must-have techniques include making your website known on every piece of literature you produce that contains your name and contact information. This includes business cards, CMAs, even signage as mentioned in the previous chapter. Mention the URL of your web presence in every medium possible.

The Technology Touch
Not only making your website known to those you come into contact with important, but let's consider those who may find you through the Internet. This is a challenge many real estate agents will face; making their website "known" and accessible to those consumers "surfing" the web.

Optimization for Search Engines

For the market of real estate, a REAL ESTATE WEBOGRAPHER™ professional must be familiar in optimizing the main presence (website) for top search engine results. Such optimization of the web site may be as simple as how the site is crafted, i.e. meta-tags, or pay services using 3rd party services.

Basics: Search Engine Optimization

For websites to gain higher search engine rankings, one must consider how web pages are designed. For each page, think of one word or phrase that encapsulates the web page. We'll call this phrase the "search term". The HTML tags are for information purposes as they are used by traditional web designers. Shortly after, described are techniques for webographers.

- *The Title tag*
  <title>some title words</title>—This should be viewed as keywords describing the entire page, like an abstract describing an academic paper, however, you should limit the words in the title tag to just a few words.

- *The Description tag*
  <meta name="description" content="a nice description">—Write an intriguing description for the page and be sure to include the page's search term least once, likely twice in the body of the description. In terms of the search term, place as close to the start of the description. Think of when you write a paper, the first sentence of a paragraph introduces that paragraph. This idea is true of search terms being at the start of a description.

- *The Keywords tag*
  <meta name="keywords" content="some keywords">—Although many search engines view these tags as plain text, many still see them as distinct words representing the web page it appears on. Place relevant keywords into the tag and include the search term at the front, and a second time further down the list of keywords.

- *The H tag*
  <Hx>some heading words</Hx>—"x" is a number from 1 to 6; the biggest heading size being 1. H tags are given more weight than ordinary text. Include the target search term in H tags at least once on the page, and many times if possible. Also, place the first H tag as near to the top of the page if possible.

- *Bold text*
  <B> Some text<b>—Bold text is given more weight than ordinary text but less than H tags. As much as is reasonable, enclose the search term in bold tags when it appears on the page.

A *Webographer-friendly hosting company* will commonly have a section in their control panel prompting you for *Title*, *Description* and *Keywords* to assist in optimizing your main presence for search engines for you. No design or HTML coding required! This holds true for RapidListings.com, agent websites. The *Title*, *Description*, and *Keyword* information are prompted by RapidListings.com of the agent user. The *In-browser, web development editor*, allows for highlighting text that is of importance, like "bold", which also appeals to higher search engine rankings.

At their option, agents can pay out-of-pocket to obtain website "traffic" from some of the major search engines on the market. It's important for the agent to focus on traffic from the local area, or looking to obtain traffic on a larger scale, like nationwide.

Resources!

At Webographers.com the discussion forum on Search Engine Optimization (SEO) is a hotbed of communication on this topic. Real estate agents are sharing best practices on how to drive traffic to their agent website.

Home Town: Search Engine Optimization

Consider a local real estate agent in St. Louis, Missouri. Besides the traditional Meta-tag technique in the header of every page, i.e. <META name "keywords" content = "Real Estate, Realty, St. Louis">, here are some techniques for the web site to reach a local audience.

- *Online Yellow Pages*
  Consider a real estate professional who desires that their business web site reach a local audience. That client should understand the Internet techniques of today's users that want to find local real estate agents. Yahoo! *Yellow Pages*, for example, provides tools to locate small businesses within local vicinity. A *Yahoo! Yellow Pages Sponsored Business*, is shown as a "Sponsored Listing" and is listed at the very top of the search results page. In addition to the business name, address, phone, and a hyperlink to the company's web site.

- *Local Matching*
  Here, the real estate professional pays for "featured" search engine results based on keywords and geographic location. Overture Services Inc, a Yahoo! company, handles the advertised web site listing architec-

ture for major companies in their distribution network. Some companies in the distribution network include such CNN.com and MSN.com. Thus a client, who pays to be a featured listing, will be listed in all the search engines that are a part of Overture's distribution network. Local Match™, a service provided by Overture, is an example of users who search from within a specified geographic region who see featured results of local businesses. This is true of all the web sites in their distribution network.[34]

Nationwide: Search Engine Optimization

A broker, agency, agent, or seller may want nationwide coverage of their properties for sale. This is true of high dollar real estate, or an area that is experiencing an influx of out-of-towners.

- *Individual Search Engine Advertising*
  The real estate professional pays for targeted search engine advertising based on keywords. Google *AdWords®* is an example of paying a fee to be a featured listing when users search on Google based on a given word.[35]

- *Multiple Search Engine Advertising*
  The real estate professional pays one fee to be listed as a featured web site across multiple search engines in a given distribution network.
  - Overture provides the Precision Match™ service, which lists the web site across multiple search engines as a featured site based on keywords.

  - Homegain.com provides the BuyerLink™ service for agents and brokers who look to drive targeted traffic to their website, specifically their "MLS listings" (IDX) web page of the agent website. Here, the agent pays a flat fee to direct traffic to their website. Traffic includes consumers looking for properties on Yahoo!, Google, MSN, and over 300 partner sites including CNNfn, CitySearch, BankRate and more. Homegain has done

---

[34] Local Match™ is a trademark of Overture Services, Inc.
[35] AdWords® is a registered trademark of Google, Inc.

the work for you through strategic alliances with these industry giants and their respective real estate, property search engines.[36]

- *Article Matching*
  The real estate professional pays one fee to be listed as a featured web site across multiple web sites in a given distribution network. The caveat to this form of advertising is the web site is listed near content, such as articles on the web sites in the distribution network. Overture provides Content Match™, which links content and relevant featured web sites together. Thus, if there is an article about a recent baseball game, a featured web site that deals with "baseball", will be listed near this article.[37]

Banner Ads

Let's examine two models of using banner ads to drive customer "traffic" to a real estate agent.

- *Pay-per-Click*
  This type of advertising means that the real estate agent may place a banner ad on a website or advertised search results, but only pays when a user clicks on the linking banner or search results. The agent can pay a specified amount they allocate to the search engine for the number of "hits" per month. Google has a popular pay-per-Click program at https://adwords.google.com/

- *Pay-per-Call*
  This method, being fairly new, connects Internet users to small businesses, through a specified and track-able toll-free number. A banner ad is placed on a website, where the ad has an area for a user to enter in a personal contact phone number. The provider of the pay-per-call services establishes a phone call between the real estate professional and the user. Real Estate professionals normally pay the service provider of click-per-call only when a customer calls the number. This is the model

---

[36] *Get Buyers to Your Website with BuyerLink™* . HomeGain, Inc.. Retrieved December, 10 2005. http://www.homegain.com/broker_solutions/index_html

[37] Precision Match™ and Content Match™ are trademarks of Overture Services, Inc.

of www.Ingenio.com, a pay-per-call service provider that caters to small business. This service is great for large brokerages looking to drive human contact to their agency. Also helps to measure return-on-investment (ROI).

Link Exchange Programs

Many search engines will rank some websites higher than others for a given keyword or search phrase due to a website's "popularity". Popularity can be determined based on the number of hits the website receives in a given time-period, or also, how many websites make reference to the target website.

Let's consider two respective, agent websites for agents who reside in St Louis, MO. Let's also say that the content and words utilized on their websites are fairly similar. However, one agent has his website referenced (hyperlinked) on 80 other websites, whereas the other agent has 2 other websites referencing his. In this scenario, the website referenced by many sites will likely have higher search engine rankings than the other agent's website.

Link-sharing programs are a network of agent's across the United States who make a point to reference each other's websites. This concept looks to drive higher search engine rankings by raising an agent's website appeared "popularity". RapdiListings.com agent websites include link-sharing capabilities; to make reference of other RapidLisrings.com, agent websites. There are many websites that promote hyperlink sharing of real estate professionals. Feel free to "surf the web" and see if there is a network that you may be interested in.

Resources!

A few examples of link-sharing websites that provide a community of real estate professionals looking to reference each other's website include:

- RealEstateLinkExchange.com
- LinkRE.com
- AvalancheRealEstate.com.
- And many others!

Chapter 7

Single-Property Websites

Yes, it's true that an agent website establishes a professional image around the good name of an agent. It presents not only the agent's current listings and value-added resources to site-visitors, but establishes a "store-front" to potential or current clients. But what in the agent's web presence really shows-off what they do for the seller?

What if you could show sellers that you have built a comprehensive web site dedicated solely to the marketing of their property? The URL (i.e. www.123AnySt.com) would match their home's address, making it easy to remember, and the whole site would be accessible by the public just hours after you get the listing.

Single-property websites (i.e. www.1410hollyst.com) are the latest craze amongst "power", top-producing real estate agents; giving each listed property its own website! They visibly show the agent's listing efforts being performed on behalf of the seller. The use of single-property websites gives credence and worth to the commission an agent earns when assisting with the sale of a property.

One Property, One Website

To some agents, a website dedicated to one property may sound over-the-top, lavish, and indicative of many hours of strenuous work. What if the development of such a website took 15 minutes? Whereas the seller is given the perception that you're selling efforts on their behalf were over-the-top, lavish and appear strenuous!

Simply stated, single-property websites show you put the client's interests first! Single-property websites means *1-property* on *1 website*. Such property websites include photos, property descriptions, floor plans, links to virtual tours, neighborhood and area information, and much more.

Street Address
Considering the URL of a property website is commonly the street address (i.e. www.123AnySt.com), it provides for a unique and easy-to-remember web address. It implies there are no competing properties listed on the website. Its one of the most powerful selling applications that should be a part of *every* listing's marketing plan.

> **Tips!**
>
> Finding an available domain these days is competitive. The probability that a domain name like 123AnySt.com is available is very high. Thus using a domain name, like street address, makes single-property websites a powerful tool for real estate professionals and their consumers.

Figure 14: Single-property Website—powered by AgencyLogic.com[38]

Residential Real Estate

In using a single-property website for residential properties, it's important to take a detailed look at the primary characteristics of a home that can influence its sale. Along with the many factors considered by a potential buyer, its easy to see that a typical newspaper classified ad cannot portray all the great qualities of a home in detail.

We've all seen newspaper classified ads displaying homes for sale. They may have room for a 25-word text description, no pictures, just a phone number and contact name to inquire more about the home. Former trends in advertising homes for sale, don't consider those who are out of the area, like in a different town or state. These potential buyers are dependent on getting a clear mental "picture" of the home before venturing to see it in person.

A REAL ESTATE WEBOGRAPHER™ professional plays an important role in enticing potential buyers of a home for-sale, as well as attracting leasers to rental properties. It's important for a REAL ESTATE WEBOGRAPHER™ professional to understand all the characteristics of a home, to ensure web-applications give key support towards the sale of the home. Single-property websites, configured in 15 minutes by an agent, can display that home in the best light.

> **REAL ESTATE WEBOGRAPHER™ professional's factors of importance:** Location (access to major highways, shopping, restaurants), school system, aesthetics (landscaping, home exterior & interior), upgrades, remodeling, potential spiritual aspects (direction home faces, number of the street address, "Feng Shui" friendly), bedroom & bathroom quantities, sq ft, room layouts, date built, price.

> **Web site Features:** Text based descriptions, map, floorplans, exterior photos (home & yard), photos of the "staged" interior (each room), virtual tours (selected rooms), contact name, email, address, phone and street address, access to documentation, brochure (pdf format).

Advertising Single-property Websites

Many agents may list their properties in publications, newspapers, free magazines, etc. Such mediums have little room for a description, or even a possible photo. Consider a small description in a newspaper ad like, "4 bedrooms, 2.5 baths, and 2,500 sq ft". This information is insufficient for local buyers and especially for potential out-of-town buyers.

---

[38] *Featured Client* . Network Earth, Inc.. Retrieved December, 10 2005. Reprinted with permission. http://www.AgencyLogic.com/

Sellers may also use online resources like Realtor.com or the popular Loopnet.com for commercial listings. Their listing is *one* among *many*. Sometimes, the webpage and allowable content for a given property on these online classified web sites may not be enough.

Real Estate Classifieds

A REAL ESTATE WEBOGRAPHER™ professional must be aware of advertising outlets that support your listing process. More importantly, it's important to note the benefits behind placing the URL of a single-property website anywhere you advertise property listings. There are many outlets commonly used in presenting a property during the listing process. Let's first look at your local newspaper.

Hard-Copy Newspapers
Consider the newspaper ad, as shown in Figure 15 below, which advertises a home for sale. This ad is a typical example of a basic home description found in the classified section of any newspaper across the nation.

A potential buyer is given enough information to possibly inquire more by telephone, and eventually view the home in-person. After reading this ad, a potential buyer is left to mentally visualize the home, since there is no space in the ad for photos, floor plans, etc. The seller is limited to the brief, property description, with the agent having high hopes that some interest is generated from a small, "text-based" ad.

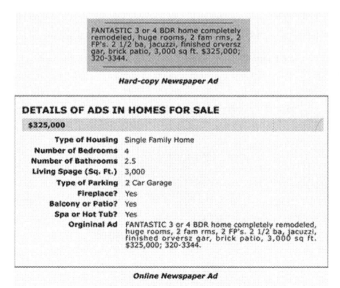

Figure 15: Residential Classified Ad—Hardcopy Newspaper & Online

Virtual Newspapers

Consider the same home as advertised in a hardcopy newspaper, but in a sample online version, also shown in Figure 15. Many newspapers have online versions of their paper, where search through real estate classifieds is possible. Consider online versions of newspapers that handle a large amount of real estate classifieds for a specific geographic region. Typically, such an online portal charges agents for postings. There is a normally a search engine associated with online classifieds, for buyers to narrow down search results based on property characteristics.

Some additional information is found from the online search, but not much more than the hardcopy, newspaper version as found in the example above. No "visuals" are provided to potential buyers due to the limitations of the provided advertising medium.

This capability is excellent for out-of-town buyers who'd like information on homes before visiting the area. But many of the same problems exist as the "hard-copy" newspapers. However, some newspapers, serving larger metropolitan areas, have included more agent features to visually enhance the property listings found in the online classified ads.

Enhanced Advertising

Now, let's consider this same agent is utilizing a single-property website for this property. How can this agent effectively use this resource where he or she may be advertising the property? As stated previously, it's important the agent place the URL of the single-property website (i.e. www.123AnySt.com), in any advertising outlet discussing the property.

Physical Newspapers

Consider the same home as discussed above. However the text-based description is appended with the URL of the web site www.123AnySt.com.

Virtual Newspapers

Consider the same home as discussed above, but the text-based description found in the on-line search is appended with the URL of the single-property website of www.123.AnySt.com.

Whether potential buyers find this home in the newspaper or on-line, they are intrigued that the current owners have a web site unique to their home. Suddenly, potential buyers will place a focus on this property. This makes this home a stand-out property. In this example, the domain name (123AnySt.com) for this site is the street address of the property for sale.

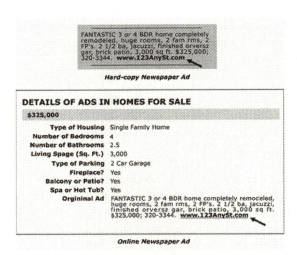

Figure 16: Enhanced Residential Classified Ad—
Hardcopy Newspaper & Online

This particular URL of www.123AnySt.com gives the notion that the site is dedicated just for this property and no other competing homes are listed. Secondly, potential buyers know they'll be able to access more information about this property without physically visiting the home.

The site, www.123AnySt.com, gives this property the ability to provide visual photos, floor plans and other in-depth details. This is extremely beneficial to the seller if potential buyers see this link in the on-line classified ad, because Internet access is already assumed available. This single-property website is personal to one property and contains no other competing properties.

Signage

Typically, real estate agents will have signs prepared for showcase outside a listed property. The content of the signs don't change and are normally fixed. However, sign riders allow real estate agents flexibility to describe the property. Sign riders are "swappable". Some sign riders an agent may have in inventory include "Large Lot", "Kitchen Upgrades", "Swimming Pool", etc.

When utilizing single property websites, a sign-rider is an excellent place to make bold mention of the URL particular to that property as shown in Figure 17. Sign Riders are meant to "pop-out" at those driving by, or who have slowed down and stopped to take a look at the property.

Figure 17: Sign Rider with URL of Single-property Website

There are a few signage providers who will print one sign-rider and ship the same day. In addition, there are sign-rider solutions that allow you to use your home-based printer to print sign-riders as needed. Sign-riders created using a home-based solution may not be the greatest of quality, but remember they are short-lived and used for the one property during its listing duration.

"For-Sale-By-Owner" Argument

Many residential homeowners use the "For-Sale-By-Owner" method to sell property. This method does not consider using traditional agents/brokers, but more so, the do-it-yourself method. The goal of this method is to avoid losing profits by paying a commission.

For-Sale-By-Owners
FSBO method has become very popular due to companies like HomeDepot. The Do-It-Yourself store sells home-selling kits in 7 southern states for $12.95. In addition, customers get a listing in Owners.com, which claims to be the largest FSBO site, with 5 million customers. The site has an answering service that fields calls from prospective buyers.[39]

---

[39] *Fast Forward*, section 55: Most Endangered Profession. Fast Company. November, 2004 Issue, p 78.

Single-property websites give the attention back to sellers who may be thinking of selling properties on their own. Many FSBO classified websites entice potential sellers, giving the seller a property web page, inclusive of many competing listings, i.e. http://www.*Some-FSBO-Website*.com/go.asp?propertyid=112233. In landing a new client, single-property websites make an agent competitive to the lure clients considering the FSBO model.

Assisted-Selling Facts

Assisted-Selling is a business model that provides a means for sellers to pay a flat fee to an agency for assistance in selling their home. Incentives for the seller include the ability to list properties in the MLS, but act with the mindset of a FSBO.

Many franchisees providing assisted-selling services will include a unique URL to assist the seller in advertising their property. This URL for a unique property may appear something like: http://www.*SomeAssistedSelling*.com/go.asp?id=112233. Can you imagine placing that URL on a sign rider or in a newspaper ad? Its simply doesn't read well, is too long and is hard to remember.

> **Tips!**
>
> For those agents who work within assisted-selling agencies or converting FSBOs, they should look to use single-property websites with a "clean" and unique URL that gives the seller true attention and marketing power.
>
> If assisted-selling agents are "bound" to defined listing services as established by their franchise, they should ask their franchise executives to include single-property websites as a listing service.

Property Management—"For Lease"

For agents who work in agencies that support both the brokerage and property management of residential properties, or strictly property management, should consider the use of single-property websites. Real estate agencies that manage properties "for lease" usually give these properties less advertising attention than properties that are "for-sale".

Why is a single-property website (i.e. www.123AnySt.com) imperative to soliciting potential leasers? Many leasers of homes are from out-of-the-area and are currently conducting search of rental homes online. Also, their stay in-town may be short-lived, as they could be working for the government as military members or business executives whose company is facilitating a short-term stay.

Single-property websites for leased properties should be used in the same fashion as mentioned above with "for-sale" properties. How they are advertised online, in newspapers, and on signage holds true for leased properties.

Tips!

Keep "live" a single-property website for a leased-property. When a new leaser is obtained, the single-property website should be kept in your inventory. Commonly, a URL for a single-property website is registered on annual intervals. Agents should inquire with single-property website providers such as AgencyLogic.com, in keeping the associated URL for a longer life span for rentals.

Commercial Real Estate

Single-property websites truly embrace the sale of commercial properties. Medical complexes, office spaces, apartment buildings should indeed have a website of their own. Commercial real estate considers properties "for sale" or "for lease". Commercial real estate is commonly owned by companies, not so much by "individuals" as seen with home-owners. Agents, who do business with commercial real estate, must add single-property websites to their listing services.

Tips!

Single-property website providers, i.e. AgencyLogic.com, provide additional display options on single-property websites specific to a commercial property. Display options for an AgencyLogic.com PowerSite may include: "For Lease Per SF", "For Rent Per SF", "For Lease Per Month, "For Rent Per Month", "Sale Price", etc.

Commercial properties are also commonly seen advertising "For Lease", where they are simply "renting out" space for business purposes. Whether selling land, selling a property, or leasing out space, a commercial real estate agent can provide additional exposure and marketing through a single-property website. This section divides commercial properties into seven (7) main categories:

1. **Office Space**
2. **Industrial Space**
3. **Vacant Land**
4. **Shopping Center**
5. **Multi-Family Homes**
6. **Retail-Commercial Space**
7. **Hospitality Space**

Below are descriptions of each commercial property type, examples of that type, and common property features that are a concern to the REAL ESTATE WEBO-GRAPHER™ professional using single-property websites. A single-property website can establish theme, helping viewers visualize an intended purpose with the commercial property. A REAL ESTATE WEBOGRAPHER™ professional can configure a single-property website to help potential buyers/leasers visualize conducting business in that property.

Office Space

Office space assumes business usage that does not include retail. Office space sales or leasing entails many different factors like location, intent of the owner of the property, etc. It's very common to see an office complex establishing a theme. Consider a complex that includes offices for a pediatrician, dentist, chiropractor, and other medical professionals grouped together. This design is due to the company who owns the property, establishing a "medical" theme for that complex. A REAL ESTATE WEBOGRAPHER™ professional's main role is to establish a visual "theme" of the intended purpose of the commercial property; and can do so with a single-property website.

> **Common Usages of Office Space:** Business Park, Governmental, High-Tech, Institutional, Medical, Mixed Use, Net Leased, Office Building, Executive Suites, Research & Development.

REAL ESTATE WEBOGRAPHER™ professional's factors of importance: Theme (medical, research, etc), location (close to major highways, residential areas, shopping and restaurants), storage space, furbished, bathrooms, date built, building size (sq ft), lot size (acres), zoning, parking spaces, landscaping, owner/broker background (sale), owner/property manager background (lease).

For-Sale Unique Factors—Fully occupied, current renters, monthly revenue, selling price/SF, cap-rate, and down-payment information.

For-Lease Unique Factors—Rent/sq ft, divisible space (sq ft), Contiguous space (sq ft), suite/floor, management/cleaning company information.

Single-property Website: Text based descriptions, map, blueprint/floor plans, exterior photos (building and lot), photos of interior, virtual tours of selected rooms, contact information, contact form, brochure (PDF format).

Industrial Space

Industrial space may include needs of manufacturing and distribution. Such properties may be considered large and more open in space. For potential buyers and leasers, its important for a REAL ESTATE WEBOGRAPHER™ professional to know of the current amenities of the given industrial space, like heater/cooler systems, processing systems, etc that may be included.

Common Usages of Industrial Space: Cold Storage, Flex Space, Food Processing, Free-Standing, Industrial-Business Park, Manufacturing, Mixed Use, Net Leased, Research & Development, Self Storage, Truck Terminal, Warehouse Distribution

REAL ESTATE WEBOGRAPHER™ professional's factors of importance: Theme (warehouse, R&D), location (close to major highways, residential areas, shopping and restaurants), rooms/space (office areas, reception area, lunchrooms, storage), furbished, bathrooms, power, ceiling height, entry doors, date built, building size (sq ft), lot size (acres), zoning, parking spaces, loading docks, truck courts, sprinkler systems, owner/broker background (sale), owner/property manager background (lease).

For-Sale Unique Factors—Current Tenants, monthly revenue, selling price/SF, cap-rate, down-payment.

For-Lease Unique Factors—Rent/sq. ft, usage type management/cleaning company information.

Single-property Website: Text based descriptions, map, blueprint images, exterior photos (building and lot), photos of interior, virtual tours of selected spaces, contact information, contact form, brochure (PDF format).

Vacant Land

From a REAL ESTATE WEBOGRAPHER™ professional standpoint, it may first appear that selling of vacant land would not need a dedicated single-property website for listing purposes. However potential buyers are always impressed by added professionalism through detailed information before they visit the site in person. Such a "land" web site developed by a REAL ESTATE WEBOGRAPHER™ professional helps potential buyers "envision" how this land would suit their business needs using a single-property website. Note. Land is mostly sold, rarely leased.

Common Usages of Vacant Land: Agriculture, Hospitality, Industrial, Multi-Family Units, Mobile Home Park, Office, Residential (Single Family), Retail, Retail-Pad, Self Storage, Vacation/Resort, Other

REAL ESTATE WEBOGRAPHER™ professional's factors of importance: Intended theme/zoning (residential commercial), location (close to major highways, residential areas, shopping and restaurants), contour (flat, slope, hilly), depth (sea level), forestation (trees, shrubbery), lot size (acres), pricing (sq ft), utility ready (telephone, natural gas, power, water), suggested ideas for use (restaurant, bank, shopping center), owner/broker information.

Single-property Website: Text based descriptions, map, ground-level photos, aerial/satellite photos, virtual tour of the outdoor space, contact information, contact form, brochure (PDF).

Tips!

Satellite/aerial imagery is important for the sale of land. Such imagery added to a single-property website allows investors to an all-encompassing view of the property and surrounding properties.

http://earth.google.com is a source to gain satellite/aerial images of land that you may add to your single-property website.

Shopping Center

Here we consider commercial properties that lease out "sections" of the property for retail purposes, like a typical strip mall. In many cases, space is still available for lease, a single-property website for that space can generate interest to leasers. A key feature to denote on a single-property website is exploiting the main store that "anchors" the shopping center. For example, it's common to see a grocery store as the anchor of a shopping center, surrounded by smaller retail stores, restaurants, etc.

> **Common Usages of Shopping Centers**: Community Shopping Center, Fashion/Specialty, Free-Standing Store, Grocery-Anchored, Mixed Use, Neighborhood Center, Net Leased, Outlet Center, Power Center, Regional Mall, Super-Regional Center, Strip Center, Theme/Festival, Other.
>
> **REAL ESTATE WEBOGRAPHER™ professional's factors of importance**: Theme (Mall, Outlet, location (close to major highways, residential areas, shopping and restaurants), storage space, furbished, bathrooms, date built, building size (sq ft), lot size (acres), zoning, parking spaces, landscaping, primary focal business (e.g. grocery—"anchored") owner/broker background (sale), owner/property manager background (lease).
>
> > **For-Sale Unique Factors**—Fully occupied, current tenants, monthly revenue, selling price/SF, cap-rate, down-payment information.
> >
> > **For-Lease Unique Factors**—Rent/sq ft, divisible space (sq ft), Contiguous space (sq ft), suite/floor, management/cleaning company information.
> >
> > **Single-property Website**: Text based descriptions, map, ground-level photos of the property, aerial/satellite photo, virtual tours (selected stores, outdoors in "plaza" center), contact information, and contact form, brochure (PDF format).

Multi-Family Listings

Such properties are for residential purposes, but commercial in mind-set. The owner leases out "sections" of the properties for rent. A REAL ESTATE WEBO-GRAPHER™ professional may maintain a single-property websites for advertising additional space for rent or for the sale of the entire property.

Common Usages Multi-Family: Duplex/Fourplex, Government Subsidized, High-Rise, Low-Rise/Garden, Mid-Rise, Mixed Use, Mobile/Manufactured Home Park, Student Housing, Senior Living, Other.

REAL ESTATE WEBOGRAPHER™ professional's factors of importance: Theme (apartment, mobile home park), location (close to major highways, residential areas, shopping and restaurants), number of units, unit information (sq ft, bathrooms, kitchen/dining, etc), pool, clubhouse, laundry facilities, building size (sq ft), lot size (acres), parking spaces, owner/broker background (sale or unit lease).

Single-property Website: Text based descriptions, map, ground-level photos of the property, photos of "model" unit, aerial/satellite photo of property, virtual tours of rooms in a model unit, virtual tours of facilities (laundry rooms, outdoor pool, fitness center), contact information, and contact form, brochure (PDF format).

Retail-Commercial Listings

When such listings are for sale, a REAL ESTATE WEBOGRAPHER™ professional may utilize a single-property website to assist the selling of this property; adding a level of professionalism to sale or leasing of the retail property.

Common Usages Retail-Commercial: Car Wash, Convenience Store, Day Care Facility, Free Standing Building, Tavern/Bar/Nightclub, Mixed Use, Garden Center, Net Leased, Movie Theater, Parking Facility, Vehicle Related, Post Office, Restaurant, Retail-Pad, Service Station/Gas Station, Other

REAL ESTATE WEBOGRAPHER™ professional Factors of Importance: Theme (Day Care, Restaurant), location (close to major highways, residential areas, shopping and restaurants), storage space, furbished, bathrooms, date built, building size (sq ft), lot size (acres), zoning, parking spaces, landscaping, owner/broker background (sale), owner/property manager background (lease).

For-Sale Unique Factors—Fully occupied, current renters, monthly revenue, selling price/SF, cap-rate, down-payment information.

For-Lease Unique Factors—Rent/sq ft, divisible space (sq ft), Contiguous space (sq ft), suite/floor, management/cleaning company information.

Single-property Website: Text based descriptions, map, blueprint images, exterior photos (building and lot), photos of interior, virtual tours of interior, contact information, contact form, brochure (PDF format).

Hospitality Listings

Another type of commercial property, that when sold most likely through an agency, stands a better chance at being sold faster and for the desired price when utilizing a single-property website.

> **Common Usages Retail Hospitality**: Bed & Breakfast, Casino, Chalet, Convention Center, Extended Stay, Golf Course, Hostel, Hotel, Inn, Motel, Recreation Cabins, Resort, Ski & Sun, Spa, Vacation Rental(s), Other
>
> **REAL ESTATE WEBOGRAPHER™ professional's factors of importance**: Theme (hotel, casino), accommodation level (economy/limited, full service), location (close to major highways, residential areas, shopping and restaurants), number of units, unit information (sq ft, bathroom(s), kitchenette, etc), pool, lounge/bar, laundry facilities, restaurant/dining, building size (sq ft), lot size (acres), parking spaces, owner/broker background (sale or unit lease).
>
> **Single-property Website**: Text based descriptions, map, ground-level photos of the property, photos of "model" unit, aerial/satellite photo of property, virtual tours of rooms in a model unit, virtual tours of facilities (laundry rooms, outdoor pool, fitness center), contact information, and contact form, brochure (PDF format).

Real Estate Classifieds

As stated earlier, a company owning commercial properties has much to gain from selling property without real estate agency assistance. This is due to the large money exchange in the sale of the property. Many companies may list their home in publications, newspapers, free magazines, etc. Their listing is one among many. Also, it has little room for a description or a possible photo. Consider a small description in a newspaper ad like, "Office space for sale, great location". This information is insufficient, especially for potential out-of-town investors (buyer).

Physical Newspapers

Consider this ad as embedded in Figure 18 that could be found in any newspaper describing the same office space property, located on 123 Commercial St., as just mentioned previously.

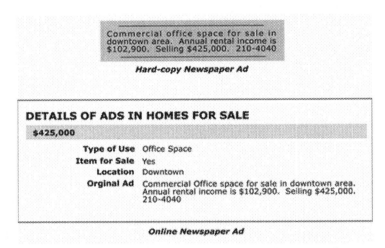

Hard-copy Newspaper Ad

DETAILS OF ADS IN HOMES FOR SALE

$425,000

Type of Use	Office Space
Item for Sale	Yes
Location	Downtown
Orginal Ad	Commercial Office space for sale in downtown area. Annual rental income is $102,900. Selling $425,000. 210-4040

Online Newspaper Ad

Figure 18: Commercial Property—Hardcopy Newspaper Ad

A potential investor is given enough information to possibly inquire more by telephone, and eventually view the property in person. What if the potential investor lives 40 minutes away, out-of-state, or out of the country? After reading this ad, a potential investor is left to mentally visualize the property, since there is no space in the ad for photos, floorplans etc. The buyer is limited to the interest generated from a small, "text-based" ad.

Virtual Newspapers

Again, many newspapers have online versions of their paper, where search through commercial real estate classifieds is possible. This capability is excellent for out-of-town buyers who'd like information on commercial properties before visiting in person. But many of the same problems exist as the "hard-copy" newspapers. Consider the same office space example as advertised, but in a sample online version as shown in Figure 18 above.

Enhanced Advertising

Physical Newspapers
Consider the same home as discussed above but appended with the mention of the web site: www.123CommercialSt.com.

Virtual Newspapers
Consider the same home as discussed above but the text-based description found in the on-line search, is appended with the mention of the web site www.123CommercialSt.com.

If a URL like www.123CommercialSt.com was utilized in various listings, potential buyers will note the professionalism and confidence in the seller or leaser of this property. On this site may include: visual photos, floor plans, benefits of location and other in-depth details. This site is specific to that given property and contains no other competing properties.

The site, www.123CommercialSt.com, gives this property the ability to provide visual photos, floor plans and other in-depth details. This is extremely beneficial to the seller if potential buyers see this link in the on-line classified ad, because Internet access is assumed. This site is personal and contains no other competing properties.

Hard-copy Newspaper Ad

Online Newspaper Ad

Figure 19: Commercial Property—Enhanced Online Newspaper Ad

Single-property websites are one of the most powerful marketing tools that put the seller's interests first. Sellers immediately see the agent's efforts put forth on their behalf. In terms of word-of-mouth referrals and repeat business, none provides that better for the agent than single-property websites.

The URL of a single-property website implies to your client and potential buyers that the website is dedicated to one property. Such a "focus" implies to buyers that the property is a stand-out worth giving attention. For seller clients, its something they want to show all their friends, driving business and new clients towards the agent!

AgencyLogic.com

AgencyLogic.com is the nation's leader in single-property websites for agents and brokers. Through an easy-to-use control panel, websites can be tailored to match the marketing goals of each unique property. By deeming AgencyLogic, as a *Webographer-friendly hosting provider*, agents and brokers can have a websites site up and running for each unique property with no HTML knowledge.

AgencyLogic includes packages for brokers (*Broker Module*) to enable all internal agents to use PowerSites at competitive, volume-pricing. Additional services such as *Custom Branding Service* which enables Brokers and Office Managers the same corporate branding is applied across PowerSites.

Tips!

Undivided attention: A Web Site for Each Listing, is a riveting article by noted columnist Michael Russer, aka "Mr. Internet" found in Realtor Magazine. The article highlights the "business" value behind AgencyLogic.com PowerSites, not to mention their ease-of-use.

http://www.realtor.org/rmomag.NSF/pages/AskMrInternet
20051031?OpenDocument.

AgencyLogic.com is a *Webographer-friendly hosting company*. As stated in the previous chapter, such a provider allows for the establishment of a website through the point & click of a mouse, and few strokes of a keyboard. No programming or HTML knowledge is required to create an AgencyLogic.com PowerSite. The web-based application includes an easy-to-use control panel accessed via Microsoft Internet Explorer, allowing agents to build custom web pages from any PC with Internet access."

Activity!

At Webographers.com, candidates of the REAL ESTATE WEBOGRAPHER™ certification work with their inclusive AgencyLogic PowerSite™ account. Candidates create a single-property website of their own, to highlight ease-of-use to best envision their powerful marketing value on behalf of the seller.

Sales Information

For sales information on establishing agent or broker PowerSite accounts, visit www.AgencyLogic.com, call toll free (888) 201-5160 or by email at info@AgencyLogic.com.

Below is an all-inclusive summary of the features of AgencyLogic.com's PowerSites (and the list keeps growing).[40]

---

[40] *PowerSites: Detailed Feature List*. AgencyLogic.com. Retrieved December, 10 2005. Reprinted with permission. http://www.agencylogic.com/wfPSFeatures.aspx

AgencyLogic.com PowerSite™ Features	
Property Information	
Product Feature	**Description**
Full-sized Homepage Photo	Makes each listing a showcase property
Full Property Descriptions	Up to 3000 characters of text (that's around 600 words!)
Detailed Property Information	Bed/bath, sq. ft., area, type and year built
"Important Message"	A highly visible and targeted message on the homepage
Detailed Room Descriptions	Write a paragraph about each room, local schools, the garden and more ... the possibilities are virtually limitless
Photo Gallery	Up to 50 photos, automatically optimized for viewing
Photo Tour	The ability to create your own elegant photo tour, with titles and descriptions
Property-Related Documents	Freely downloadable or password protected
Floor Plans	Upload floorplans in a variety of formats
Area Schools	Reports specific to the home's location
Virtual Tours	Upload a link to a virtual tour
Property Map	And the ability to obtain driving directions
Open House Page	List your open house dates
Print Brochure	Includes all property information
Mortgage Calculator	Automatically enters the listed price
Professional Design	Elegant, top-notch designs that showcase your listing
Contact Tools	
Product Feature	**Description**
Schedule a viewing	Lets your prospect schedule viewings automatically and simply
Email a friend	We tell you who emailed whom about what property
Contact Me Page	Easy way to reach you
Track Document Downloads	Capture users' names and email addresses

Agent Marketing	
Product Feature	**Description**
Contacts and Qualifications	Listed on top of the page
Contact Numbers	On every page
Agent Photo	On every page
Your Website	Links back to your personal or corporate Website
Company Logo	On every page

Virtual Tours

A popular resource with property listings on an agent web site, single-property website, and online classifieds like Realtor.com is a virtual tour of a property or space. It gives users an interactive view of a property and gives the sense of realism that they are touring a kitchen, master bedroom, or backyard "in-person". A potential buyer, who is even out-of-the-town, can view virtual tours from their own home, on their own time.

Figure 20: Published Virtual Tour—RealBiz360.com [41]

---

[41] *Virtual Tour Demo 4*. RealBiz360.com. Retrieved December 10, 2005. Reprinted with permission. http://www.realbiz360.com/products/sample_tours.htm

Buyers visiting an area from out-of-the-area looking for a property may be on a limited visiting schedule. Such buyers want to plan a trip and view homes in advance. The same holds true for local buyers, looking to plan a Saturday for open-house visits. Thus a REAL ESTATE WEBOGRAPHER™ professional should look to include more than just still images and use of virtual tours.

Eighty-four percent of home buyers say photos and detailed property descriptions are the most important features when searching online for homes—*followed closely by virtual tours*[42]. Thus, all agents must engage this powerful listing and marketing tool. In addition, making virtual tours yourself or having your assistant create virtual tours is a lot easier than what you may think.

Create a Virtual Tour Yourself

A Virtual Tour is a collection of scenes (i.e. kitchen, master bedroom, living room, etc). Each scene is one large panoramic image (which can include many stitched images). The panoramic tour viewer found in the browser, gives a sense of realism or feeling of a "walk-thru" of each scene.

Overarching Practices

There are a couple methods to obtain a 360-degree or partial panoramic image, which is then made into a spinning (rotating) virtual tour:

1. One-shot image
2. Stitched images

One-shot Image
The *one-shot image* process includes the use of a panorama lens attachment to add to many common digital cameras currently on the market. This process, takes 1 photo for full panoramic image, to later be presented as a scene in a virtual tour.

Stitched Images
The *stitched image* process includes the use of the everyday digital camera, as-is. Although purchase and use of a tripod and pano-head (later described in this chapter) can be implemented, this process speaks to the everyday real estate

---

[42] 2004 National Association of REALTORS® Profile of Home Buyers and Sellers.

agent, using tools they likely have in-stock, such as a digital camera. This process includes (a) photo-taking, (b) stitching the images together, and (c) outputting the scene for view in a tour. Virtual tour providers have made steps "b" and "c" easy. The hardest part is taking quality photos, that's it.

"Pay-per-Tour" vs "Monthly Subscription"
Given the two scenarios to complete self-serve virtual tours, service providers normally present one of two payment options of (a) per tour or (b) a flat-rate monthly subscription. Although a decision based on the feelings of the individual agent, such a determination is based on expected volume of listed homes per month. With an expected volume in-hand, one can then determine the most cost-effective solution.

Selecting a Service Provider
When making the decision to self-serve virtual tour creation or possibly through a vendor who does the tours for you, simply look at any "sample" virtual tours. Can you zoom in each scene? Do the scenes look distorted or have a fish-eye (curved) appearance? Simply take a look at samples located on the provider's website.

Although various methods to produce and self-serve quality virtual tours exist, National Institute of Webographers focuses on the above stated process as the *commonly used practice*.

Commonly Used Practice

The process of creating a virtual tour of a given space includes 3 major steps: (1) Digital Photo-taking, (2) Stitching, and (3) Outputting. Digital photo-taking requires taking still images of a space while horizontally rotating the digital camera from a fixed position. Stitching includes the use of software to find matching points between photos and merging that collection photos into one larger photo (a panorama). Outputting refers to publishing the scene for viewing from a specified web-page and inclusive tour viewer.

Sound complicated? Rest assured that virtual tour providers, like RealBiz360.com, provide solutions to make the creation of virtual tours easy. The hard part is for the agent to go and take pictures!

How to Shoot for Virtual Tours

Self-serve virtual tours begin with quality photo-taking in digital format. Minimal requirements include a digital camera and memory card reader or USB cable to transfer photo images to a personal computer (PC) from the digital camera. Optional requirements may include a tripod, tripod head (pano-head) to keep the camera level (if you don't have a steady "hand"), and a wide-angle lens. Chapter 15 describes available hardware in today's market to best take digital photos for virtual tours.

Techniques in taking photos for virtual tours are best described in *Agent's User Guide: Shooting You Photos* written by Marsha Scharf of RealBiz360.com. This comprehensive guide is generalize-able to quality photo-taking techniques, but speaks to the RealBiz360 virtual tour marketing system, found exclusively at www.RealBiz360.com. The following steps below are written in the spirit of this well-written guide.

1. **"Open-House" Mindset**
   Virtual Tours help to replace the "physical" walk-thru of a Sunday afternoon, open-house. Thus, the photo-taking for a virtual tour, like an open-house, should be "staged" accordingly. Be sure to assist the seller in preparing their home for the photo-shoot, just like any open-house. Thus a pre-visit, by an agent or an assistant may be required, to ensure the home is neat and free of clutter during the day of photo-taking.

2. **Analyze The Room**
   The size, shape and lighting of a room play an important role in the quality of photos for virtual tours. For small spaces, photo-taking best occur by standing in the corner of a room. Larger spaces entail being in the middle of a room.

   Some tips before taking photos in a room include:

   a. Take photos in the evening if-possible, to avoid bright sunlight coming through windows.

   b. Make sure a room is well-lit. Turn on all accessible lights.

   c. For dim rooms, where additional lighting is not available, use your camera's flash with every shot.

   d. For adjacent rooms, open their respective doors, and make sure those rooms are also well-lit.

3. **Digital Camera Only**

Can photo-taking for a virtual tour be done with just a digital camera, with no tripod or tripod-head (pano-head)? The answer is yes, but it takes a steady-hand.

Camera Positioning

In short, one must hold the camera firmly. It's best to hold the camera vertically (portrait-mode), to obtain as much of the floor-to-ceiling space as possible. For small rooms, you may look to obtain a partial panorama. This means you position yourself in the corner of a room or a doorway, taking as many photos as required. The tour you will finally produce won't turn a full 360-degree, but more of a 180-degree panorama.

First Photo

Pick a distinct feature in the room that helps you remember where you started taking your photos, this could be the edge of a door or window.

Turn Your Body Evenly

With each photo, make sure the camera is the same distance from ground. To avoid parallax (image mismatch) don't tilt the camera forward or backwards. Keep the camera always parallel (even) with the floor when photo-taking. Don't "drop" your hands as you rotate your body from a fixed position. This keeps the camera on an even horizontal line while rotating.

Each Photo

Take photos sequentially around the room. Ensure each photo has 20%-50% overlap. If rotating your body to the left for example, a window that may appear in the left-side of one photo will be in the right-side of the next photo. This is because you're ensuring overlap between photos.

All-in-all, for a 360-degree panorama, you're looking at taking anywhere between 12-18 photos, with the last photo overlapping with the first photo.

Tips!

For outdoor photo-taking, look at times in the day when the sun is highest in the sky (around 12 noon) to avoid deep pockets of shadows.

Figure 21: Photo-taking with Tripod, Tripod Head, and Digital Camera[43]

Resource!

VirtualTourWebStore.com (a "sister" service of RealBiz360.com) provides digital cameras, tripods, tripod heads, carrying cases and bags for professional photo-takers. In addition, they provide "bundled" packages for one to purchase an "integrated" kit; perfect for yourself or an assistant.

As shown in Figure 21 above, a tripod and a pano-head are not required unless phototaking "by-hand" is unsuccessful. Again, when using just a digital camera, its takes a steady and level hand throughout photo-taking.

Produce & Publish the Virtual Tour

After photo-taking has been performed, let's examine the next couple steps in *stitching* the photos into one panorama, followed by *outputting* the scenes for full virtual tour viewing.

Stitching the Photos
Self-serve providers have the core process of stitching scenes and providing upload to a user account on the provider's platform. There is a universal

---

[43] *Agent's User Guide: Shooting You Photos.* Marsha Scharf. RealBiz360 Reprinted with permission. http://www.realbiz360.com/pdfs/RBshootingyourphotos.pdf.

process performed by users of self-serve virtual tour providers that use a stitching technology.

Case-Study

For case-study purposes, we'll look at Realbiz360.com's Tourbuilder as an example for this process.

RealBiz360.com provides all of its users with TourBuilder, an application built from IseeMedia's well-known stitching application called Photovista. Panoguide.com, the "leading guide to panoramic photography and object movies", has users on its discussion forums constantly stating Photovista is one of the easier stitching application for virtual tours and great for beginners. From that Photovista technology of IseeMedia, has arisen an easy-to-use stitching program, TourBuilder, which automatically uploads your stitched photos to your RealBiz360 user account.

Let's recall that in the "How to Shoot for Virtual Tours" section in this chapter, two major things occurred:

1. Photos were taken in portrait mode (camera held vertically) to get as much of the floor-to-ceiling space.

2. Photos were taken sequentially in a circular fashion.

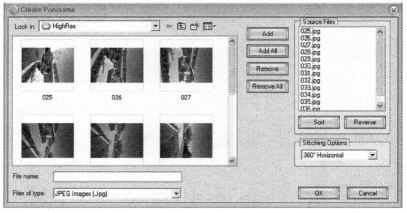

Figure 22: TourBuilder Photo Selection—RealBiz360.com[44]

---

[44] *Create Panorama—TourBuilder Application.* RealBiz360/ IseeMedia Reprinted with permission.

As one can see in Figure 22, the images selected in TourBuilder appear on their "side", due to taking the photos in Portrait mode. After selecting the desired photos from your computer, TourBuilder can rotate all the images back to landscape mode for stitching.

Secondly, in the "Source Files" box shown in the TourBuilder screenshot in Figure 22, the images are numbered 025.jpg, 026.jpg, 027.jpg. It's important to note that digital cameras name their pictures in sequential order, whether alphabetically or numerically. TourBuilder assumes that adjacent photos are "named" in a sequential fashion, like '1.jpg', '2.jpg', '3.jpg'.

After stitching, TourBuilder will automatically upload a stitched panorama (a scene) to the respective property within a RealBiz360 account, all with the click of a mouse!

File Size
It's important for a REAL ESTATE WEBOGRAPHER™ professional to understand the file size of stitched, panoramic image. Simply stated, a panoramic image consists of many stitched JPEG's or GIF's. Just like a high resolution photo on the web can determine upload/download time, imagine many stitched images of a panoramic at high resolution. This can be a very large file. Be sure to test the upload time of a tour using TourBuilder (or your preferred application) and download time of a completed virtual tour. One should test upload/download speeds on a slow connection, i.e. 56k modem, in addition to DSL/Cable/T1 connections to ensure proper usability (download) of your virtual tour.

To reduce the overall size of a stitched panorama, a REAL ESTATE WEBOGRAPHER™ has two options:

1. Reduce Camera Settings—set the camera to take low resolution photos.

2. Compression—After selecting and rotating photos, stitching software normally has options to create a stitched panorama at a desired resolution. As with the case-study of TourBuilder, the three options that allow one to be mindful of file size include:
   a. *Standard Resolution*—Fastest Upload Speed
   b. *Enhanced Resolution*—Faster Upload Speed
   c. *Camera Resolution*—Variable Upload Speed

Prepare the Virtual Tour

After upload of the stitched panorama (scene) to a RealBiz360 account (the case-study), a user can apply descriptions, music and additional features to the room or space within the virtual tour.

It's important to note that a virtual tour is a composite of many panoramas or scenes (i.e. kitchen, master bedroom, living room) with descriptions of each scene keyed in by the agent or an assistant. In Figure 23 below, displays how you can apply a "title" and description to a panoramic image, and even apply music to the whole virtual tour.

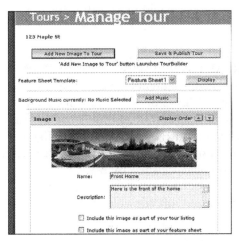

Figure 23: Managing the Virtual Tour—RealBiz360.com[45]

Publish the Tour

Once all panoramas (scenes), descriptions and features have been applied to the all-inclusive virtual tour, its now time to publish the virtual tour. RealBiz360.com, the well-known leader in virtual tour marketing provides three options to publish tours as shown in Figure 24. The importance behind the functionality of RealBiz360.com is that all the steps to create a virtual tour are done once! Once created, you may at *any time* and *any number of occurrences*, select the desired way to publish the virtual tour. Again, the virtual tours created at RealBiz360.com are reusable as with many virtual providers on the market!

---

[45] *Manage Tour.* RealBiz360.com. http://builder.realbiz360.com/vtagent_admin.asp Reprinted with permission.

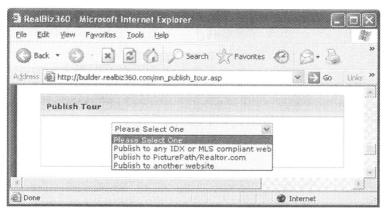

Figure 24: Publishing the Virtual Tour—RealBiz360.com[46]

1. *Publish to any IDX or MLS compliant website.*
This option allows for a unique URL to be generated, to hyperlink with your MLS listing. However the template this virtual tour resides does not include the agents, picture and contact information (as required by many MLS's)

2. *Publish to PicturePath/Realtor.com*
With the importance of using a PicturePath™ provider such as RealBiz360.com, tours can be seamlessly published to Realtor.com for a nominal fee. The virtual tour resides on a template you select, which can include your agent picture and contact information.

3. *Publish to Another Website*
This option allows for a unique URL to be generated, to hyperlink the hosted tour from any website. The virtual tour resides on a template you select, which can include your agent picture and contact information. As described earlier in the chapter the tour can be hyperlinked from your agent website, single property website, or online classifieds like *Yahoo! Real Estate* or your local newspaper's online classifieds!

---

[46] *Publish Tour.* RealBiz360.com. http://builder.realbiz360.com/mn_publish_tour.asp. Reprinted with permission.

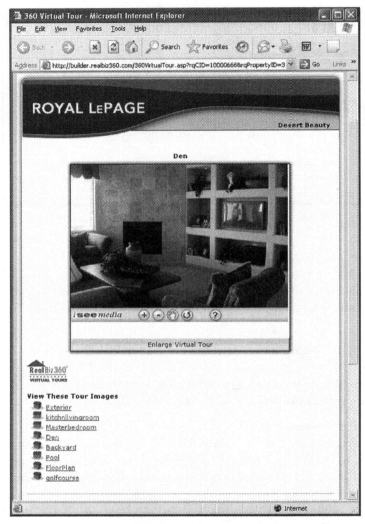

Figure 25: Published Virtual Tour—RealBiz360.com [47]

---

[47] *Virtual Tour Demo 4*. RealBiz360.com. Retrieved December 10, 2005. Reprinted with permission.   http://www.realbiz360.com/products/sample_tours.htm

Resource!

Panoguide.com is a community-centric website that provides discussion forums for Q&A on software, hardware (tripods, pano-heads, lenses, etc.) and other topics in panoramic imaging and virtual tour creation for beginners to experts.

Webographers.com and its Virtual Tour discussion forum is a hotbed of communication on panoramic imaging and virtual tour creation specific to the real estate market. REAL ESTATE WEBOGRAPHER™ professionals discuss best practices in creating virtual tours and their use as a marketing and listing resource.

Copy to CD, floppy, email, etc

Many agents desire the ability to copy virtual tours to CD, floppy or even email virtual tours. When selecting any of the 3 publishing methods at RealBiz360.com a user may make a copy of the tour for any desired use.

In terms of providing sales materials to potential buyers, nothing is more impressive than supplying such interactive media on the property.

Place Virtual Tours Everywhere

In chapter 4, *Webography Process*, the agent establishes their requirements, selects service providers and establishes a seamless web presence. In terms of a seamless web presence, virtual tours can be referenced in many outlets found in an agent's web presence. Once created, virtual tours are a "reusable" resource.

In this section, we will focus on displaying your virtual tours in three key areas:

1. Agent Website
2. Single Property Website
3. Online Classifieds

Agent Website

With every property listing on your agent website should be an inclusive virtual tour. For agents self-serving the creation of virtual tours, one must look to a provider, like a Realbiz360.com, for extensive publishing capabilities. Let's not forget, virtual tours are reusable because they are commonly hosted as a sole-entity. This simply means the virtual tour is accessible from a unique URL, like http://www.<virtual-tours>.com/?id=112233. This enables an agent to provide links to a given virtual tour, anywhere the respective property is listed.

When adding, editing or modifying a property listing in an agent website, as powered by RapidListings.com for example, the capability to add a hyperlink to a "hosted" virtual tour is made available. When the website and property listing is viewed by site visitors, a "spinning house" makes known a virtual tour is accessible for viewing as found with RapidListings.com

In Figure 26 below, one can see how the URL of the hosted virtual tour is added to a property listing on an agent's website. This is commonly done within the control panel of your agent website as supplied by your *Webographer-friendly hosting company*.

Figure 27, also shown on the next page, details how that tour is seamlessly accessed by site visitors from the property listing's web page of the agent website; by clicking on the "spinning house" associated with that property listing.

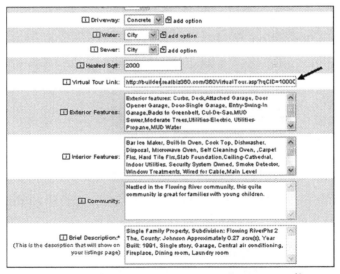

Figure 26: Edit Property Listing—RapidListings.com[48]

---

[48] *Manage Listings.* RapidListings.com. Retrieved December 10, 2005. Reprinted with permission. http://www.agentedit.com/listing_edit.asp?

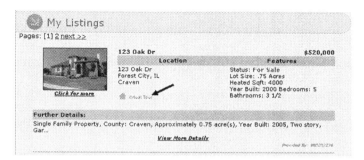

Figure 27: Property Listing with Virtual Tour—RapidListings.com[49]

Single Property Websites

Given that the tour is hosted at a unique URL, that same virtual tour can be hyperlinked from a single-property website as well. Simply stated, you make reference to that hosted virtual tour within the control panel of your AgencyLogic.com, PowerSite account.

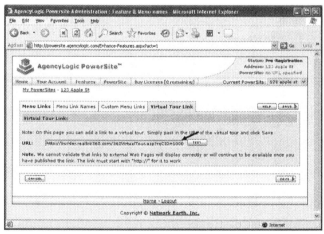

Figure 28: Edit Property Listing—AgencyLogic PowerSite [50]

---

[49] *Preview Agent Website.* RapidListings.com. Retrieved December 10, 2005. Reprinted with permission. http://www.agentedit.com.

[50] *Virtual TourLink.* AgencyLogic.com. Retrieved December 10, 2005. Reprinted with permission. http://agencylogic.powersite.com/Enhance-Features.aspx?act=1
Note: The Enhamced Features list is available only to logged on users

Online Classifieds

Online Classifieds such as *Yahoo! Real* Estate or your local newspaper (online version) have a similar process in incorporating virtual tours with a given property listing. Commonly, such online classifieds allow for a user to enter in the URL of a hosted virtual tour with every property listing.

However, the most prized online classified outlet in existence is Realtor.com. Realtor.com and it's popularity holds true for both the real estate professional as a preferred listing resource, and from the perspective of site visitors looking for the perfect home. Amongst all the websites that agents could place their property listings, Realtor.com is ranked #1 amongst surveyed Realtors at a whopping 84% usage by all Realtors®.[51]

Realtor.com and Virtual Tours

There are huge advantages for the agent to have a virtual tour with every property listing found at Realtor.com. For potential home-buyers searching on Realtor.com, such listings that have an inclusive virtual tour are marked with a "Red Spinning Box".

In fact, users can limit their search results to those listings that must have tours. Many will only view properties that have virtual tours. In addition to Realtor.com hosting the tours, Homestore has established the Virtual Tour Distribution Network™ (VTDN)[52] which includes many national, real estate portals as listed below.

Virtual Tour Distribution Network™ (VTDN)

- Homes.com
- Homeseekers.com
- ColdwellBanker.com
- Century21.com
- ERA.com

---

[51] *Web Sites Where Realtors® Place Their Listings..* 2004 National Association of REALTORS®: Technology Impact Survey, p 23.

[52] REALTOR.com®, hometour360°™ and the Virtual Tour Distribution Network™ are trademarks of Move, Inc. (formerly Homestore, Inc.)

- PrudentialRealEstate.com
- RealtyExecutives.com
- Realtor.com
- REMAX.com
- Hundreds of regional and local broker and office sites; and numerous affiliated agents

For many agents looking to incorporate virtual tours as a part of their listing process on Realtor.com and the Virtual Tour Distribution Network (VTDN), there are two major virtual programs to consider: PicturePath™ and Hometour360™ program (i.e. *self-serve* or *out-source*)

PicturePath™ program

PicturePath™ virtual tours are hosted on the service provider's servers (like RealBiz360/IseeMedia) and not Realtor.com.[53] This allows for companies like RealBiz360.com to have its users create tours once and "publish" to various outlets, like Realtor.com, or to a unique URL for their local newspaper (online), agent website, or single-property website. PicturePath™ providers allow flexibility in where you want virtual tours to appear, because companies like RealBiz360.com are hosting the tour, not Homestore Inc. (manager and co-owner of Realtor.com).

PicturePath™ providers will take a virtual tour and upload information about the tour to the property listing on Realtor.com. Essentially, the virtual tour is "matched" to the property listing found on Realtor.com. Agents must sign the REALTOR.com® *Virtual Tour Linking Agreement* to use PicturePath and add virtual tours to listings.

---

[53] PicturePath™ and Hometour360 are trademarks of Move, Inc (formerly Homestore, Inc.)

PicturePath™ providers (like RealBiz360) are charged a per tour and distribution fee which is published in the PicturePath Rate Card. Thus, any PicturePath™ provider must in-turn, charge its customers for upload of virtual tours to Realtor.com.

Advantage to PicturePath™ Providers

1. As a customer of a PicturePath™ provider, you know that no intensive efforts are required on your part to "get" your tours on Realtor.com. PicturePath™ providers have met all requirements and guidelines to publish tours to Realtor.com and the Virtual Tour Distribution Network (VTDN).

2. PicturePath providers host their own tours. Agents can hyperlink the property tours on *other* web sites, like individual agent web sites, single-property websites, online classifieds like Yahoo! *Real Estate*, and any other medium that may not be part of the VTDN.

Hometour360™ program

Hometour360™ program was established for agents and brokers to have outside help (local photographers as resellers of HomeTour360) to create and post virtual tours to Realtor.com. The program includes "a network of 'authorized service providers,' photographers who can deliver the full-service HomeTour360 solution to agents and brokers in local markets". [54]

Advantage to the Hometour360™ program

1. As a customer of the Hometour360™ program someone else is taking care of the photo-taking, virtual tour creation, and upload of tours to Realtor.com.

2. Virtual Tours are published to Realtor.com and the Virtual Tour Distribution Network (VTDN).

Disadvantage to the Hometour360™ program

1. Sometimes, scheduling conflicts occur with photo-taking opportunities between the (a) resident-owner of a property and the (b) *Hometour360*, authorized service-provider.

2. It may take up-to 3-4 days to finally "obtain" the completed virtual tour. Distribution Network.

---

[54] *Homestore Revamps Its Virtual Tour Business.* Blanche Evans. Realty Times. Published: May 17, 2002. Obtained December 10, 2005. http://realtytimes.com/rtapages/20020517_homsvt.htm.

RealBiz360.com

RealBiz360.com provides a complete, online, virtual tour marketing solution designed specifically for the Real Estate professional—that allows unlimited tours for one affordable subscription. This self-serve, virtual tour solution provides tour upload to Realtor.com as a PicturePath™ provider, MLS/IDX, or to any website. Realbiz360 uses photo-stitching technology stemming from iseeMedia's Photovista® product line, hailed as the industry leader in virtual tour creation for nearly a decade, and constantly recommended for beginners in panoramic stitching/virtual tour creation. That power and ease-of-use is brought forth, at Realbiz360.com for agents, brokers and their assistants to self-serve their virtual tour needs.

High-Definition Virtual Tours™ (HDVT) delivers an amazing sense of realism to the real estate buyer, enabling users to zoom in and see fine details within a Wide panoramic viewer. When a user "zooms" on distinct areas of a given image in the tour, resolution is adjusted to provide clarity.

Figure 29: High Definition Virtual Tours—RealBiz360.com[55]

[55] *RealBiz360 High Definition Virtual Tour.* <u>RealBiz360—National Association of Realtor Conference Nov 10-13—Brochure</u>. RealBiz360. Reprinted with permission.

Steve Marques, vice president and managing director for Toronto-based iseeme-dia, which has partnered with RealBiz360, says:

> *"'You can see the wood grain on a dresser from across the room', of course, that can put more pressure on sellers, he joked, 'What happens if I don't have a clean house?"* [56]

Activity!

At Webographers.com, candidates of the REAL ESTATE WEBOGRA PHER™ certification work with their inclusive RealBiz360™ virtual tour, marketing account. Candidates create a virtual tour for training purposes using the well-known TourBuilder application.

Sales Information

For sales information on establishing agent or broker Realbiz360 account, visit www.RealBiz360.com, call toll free at 1.888.REALBIZ (732-5249) ext. 81 to speak with a sales representative, or email sales@realbiz360.com.

The table on below is an all-inclusive list of RealBiz360.com, virtual tour marketing solutions and features.

RealBiz360.com Features List [57]		
RealBiz360.com Solutions		
Product Feature	TourBuilder	WebCreator\Tour Builder Solution
Unlimited tour creation	X	X
Unlimited tour uploads	X	X

---

[56] *Part 1: Next generation of virtual real estate technology.* Glenn Roberts. Inman News Newsletter. http://www.inman.com/member/newsletter/0623vir/story.aspx?ID=46732. Published June 21 2005, Obtained December 10, 2005.

[57] *Virtual Tour Marketing Solutons.* RealBiz360.com. Obtained December 10, 2005. http://www.realbiz360.com/products/realbizsolutions.htm. Reprinted with permission.

Stitching software	X	X
Publish tours to marketing websites	X	X
Email virtual tours	X	X
Customized tour templates	X	X
Add pictures to your tours	X	X
Contact information	X	X
Upload tour banner	X	X
Web based tour management	X	X
Tours technical support	X	X
Online Tutorials	X	X
Unlimited webpages		X
Edit website using web browser		X
Company branded website templates		X
Upload website banner		X
Industry tools		X
Website statistics		X
Step by step wizard		X
Website hosting		X
Website technical support		X
Free domain name		X

RealBiz360 TourBuilder	
X	No Special Equipment Required
X	Easy-to-Use Interface
X	Powerful 360° "Stitching Engine"
X	Preview Feature to View your Tours before Uploading
X	Create Still Images or 360° Panoramas
X	Easy Upload to your TourBuilder Online Account
X	Easy to Manage your Tour, Create, Edit, Delete and Publish
X	Include up to Eight Tours for Each Property
X	Free Viewer Templates
X	Low Monthly Fee—Unlimited Tours

X	Tours are Professionally Hosted Within a Secure Dedicated Technical Facility that is Insured with Power Backups
X	Includes Interactive Viewing Feature—Ability to Zoom in for a Closer Look—HDVT

PART FOUR:
Extend Your Web Presence

Chapter 9: MLS IDX VOW ILD Technologies

Chapter 10: Neighborhood Search

Chapter 11: CMA's and AVM's

Chapter

9

MLS IDX VOW ILD
Technologies

Recently, the National Association of Realtors® has looked to change policies for Internet Data eXchange (IDX) and Virtual Office Websites (VOW). IDX allows brokers within an MLS territory to include each others' listings on their websites. Virtual Office Websites (VOW) is a business model that distinguishes (1) MLS participants' property displays on the Internet from (2) the displays governed by the IDX policies or participating brokers.

Both the MLS IDX and VOW policies look to be replaced with the single policy known as the Internet Listing Display (ILD). This policy looks to create greater equality in the showcase of MLS listings on any website.

Although the new Internet Listing Display (ILD) policy looked to be in effect for July 1, 2006, this chapter discusses both IDX and VOW policies, to show evolution to this new policy and what it means to a broker or agent's website.

Multiple Listing Service (MLS) Overview

The Multiple Listing Service, or MLS is service created and run by real estate professionals which gathers all of the property listings into a single place so that purchasers may review all available properties from one source. MLS databases are local to a given region and are largely managed and owned by Realtor® associations. The MLS also deals with commission splitting and other relations between brokers and agents. All in all, the MLS is a listing service, with a given technology provider "managing" a given geographical region. However, many rules on usage of the MLS are defined by participating brokers of that given region.

MLS RETS Compliance
Local MLS databases may seem like "lone" islands of property-data specific to an area. A few years ago, every local MLS was implementing their system in different ways, where MLS systems could not "talk" to each other and were incompatible with one another.

However, all MLS systems should now adhere to the Real Estate Transaction Standard (RETS). RETS is a "common language spoken by systems that handle real estate information, such as multiple listing services."[58] Developers and software providers for IDX services for example, must adhere to working with this standard·

The goal of RETS was, "to provide a single interface standard that spurs the development of new and innovative tools for real estate companies and agents to use."[59] In short, 3rd party applications and many off-the-shelve products that work with the MLS, like Top-Producer, must be RETS-compliant.

MLS Input & Access Methods
In the 2005 MLS Technology survey, performed by the Center of REALTOR® Technology (CRT), had 524 respondents discuss overall usage of the MLS. 100% of the respondents state that Internet access is available to their MLS. That's encouraging to know that the MLS supports the objectives of a REAL ESTATE WEBOGRAPHER™ professional: ability to perform real estate objectives over the Internet, even when mobile. 56% of agents and brokers report their MLS provides this service, while 37% actually use wireless access service.[60]

In terms of who enters in data into the MLS, 75% state that agents enter in their own listings, with 58% state that office staff has access to enter in data into the MLS,

[58] *Getting Started with RETS*. Real Estate Transactions Standard (RETS) Working Group. http://www.rets.org/userstart.html. Retrieved 05 Jan 2005,
[59] *Will The Real Estate Industry Standarize Internet Data Exchange?* Greg Herder. Realty Times. http://realtytimes.com/rtapages/20041207_idxstandardization.htm. Published: December 7, 2004. Obtained: December 10, 2005.
[60] *2005 MLS Technology Survey*. Center for REALTOR® Technology CRT. Published March 31, 2005. http://www.Realtor.org/CRT, p22.

as a support role. [61] In addition, the majority of respondents state they input listings via the Internet using a browser. Other input methods include FTP using a batch method, email to the MLS, Fax to the MLS, or a stand-alone client application.[62]

Agent's Website & MLS Listings

Incorporating local Multiple Listing Service (MLS) listings on agent or agency website is a way to advertise all local listings on an agent's website. Such incorporation keeps visitors at the website longer, keeping them from going elsewhere to find more information on local listings. With that stated, let's discuss some essential background information on the MLS, and how it pertains to an agent or agency website.

Internet Data eXchange (IDX)

IDX allows brokers within an MLS territory to include each others' listings on their websites. Brokers must have the consent of brokers within the exchange to showcase the listing of others.

Caution!

In some areas, IDX display is limited to just the brokers (agency/firm) website and cannot be displayed on an agent's personal website. Check with your broker or MLS / Association if adding IDX listings is allowed on your website!

The MLS data presented on an IDX web page is typically less detailed, and viewers of the data remain anonymous.[63] There are many positives of integrating MLS

---

[61] *2005 MLS Technology Survey.* Center for REALTOR® Technology CRT. Published March 31, 2005 http://www.Realtor.org/CRT, p9.

[62] *2005 MLS Technology Survey.* Center for REALTOR® Technology CRT. Published March 31, 2005. http://www.Realtor.org/CRT, p10.

[63] Field Guide to Virtual Offices Web sites (VOWs) and Internet Data Exchange (IDX). National Association of Realtors. http://www.realtor.org/libweb.nsf/pages/fg902 (Retrieved 25 Apr 2005).

data on agent's website. One major importance is that visitors can inquire specifics on the property and local market when they remain at the agent/agency website.

Framed IDX. Many MLS providers provide brokers and agents the ability to include a framed-webpage that has an established search tool and presentation of shared listings. The search may include such property types as Residential, Residential Rental, Land & Lots, Commercial/Industrial For-sale, Commercial Industrial for Rent, and Multiple Dwellings.

In terms of including such a resource on an agent/agency website, this entails the inclusion of a URL that identifies where this framed page is hosted. *Webographer-friendly hosting providers* normally have an area on the administrators control panel, where one can simply add the framed page by typing in the MLS IDX hyperlink (URL). In order to have access to the IDX site, an agent or broker must commonly have a valid active Agent ID, a valid website address (URL) and have their real estate license hung with a broker that has not opted out of the IDX program.[64]

> Technical Details. Although may be performed differently among different MLS providers, here is an example from the Arizona Regional Multiple Listing Service, INC (ARMLS.com) For example purposes, here's how to frame the ARMLS-IDX website:
>
> 1. *"...Verify the URL.* The agent or broker must enter their website address into the TEMPO system...this address must be a valid URL in a standard format similar to these examples: www.myrealty.com, www.maryclark.com, etc..."
>
> 2. *"...Build a Frameset.* Use the following to code the source URL for the frame responsible for holding the IDX website. Replace XX123 (below) with the 5 character alphanumeric Agent ID assigned to the agent framing the site..."
>
>   "...<framesrc=http://armlsidx.vstone.com/?agentid=12345 width=510>..."

For *Webographer-friendly hosting companies* like RapidListings.com, all you need is the URL of the IDX page to add the web page to your agent website; simple point and click!

---

[64] *How to Frame ARMLS IDX Data.* Arizona Regional Multiple Listing Service. Published: http://www.armls.com/pdfs/ARMLSAgentIDXHowTo.pdf. Published: Novemeber 15, 2002. Obtained December 10, 2005.

FTP—IDX. File-Transfer Protocol or FTP of MLS data is the transfer of raw MLS data from the MLS Provider to the hosting provider of the website. The advantage of this process is the flexibility in designing the search feature and display of search results. Many *Webographer-friendly hosting companies* have pre-built web pages that displays FTP'd IDX data on an agent or agency website. Essentially, the hosting company should do the work for you if MLS IDX can only be obtained via FTP. Be sure to ask your local MLS provider about how they accommodate MLS IDX requests.

In the 2005 MLS Technology Survey, IDX use is widespread with 25% of MLS respondents indicating that over 50% of their members use IDX.[65]

Resources!

APPENDIX B.—RapidListings.com MLS IDX Coverage, denotes the hundreds of MLS coverage areas that are readily compatible with a RapidListings.com agent or agency website. RapidListings.com can work with MLS IDX, or raw FTP'd MLS feeds.

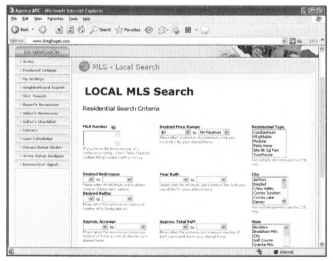

Figure 30: "Framed" MLS IDX on Agent Website—RapidListings.com[66]

---

[65] 2005 MLS Technology Survey. (2005, March 31), WAV Group in conjunction with Center for REALTOR® Technology (CRT). p34. http://www.Realtor.org/CRT.

[66] *Preview Agent Website.* RapidListings.com. Retrieved December 10, 2005. Reprinted with permission. http://www.agentedit.com/.

Virtual Office Websites (VOW)

Virtual Office Websites (VOW) was a business model that distinguishes (1) MLS participants' property displays on the Internet from (2) the displays governed by the IDX policies or participating brokers. "The primary distinguishing feature of a VOW is the requirement that visitors register by entering an email address and receive a password prior to accessing MLS listing data."[67]

The VOW business model was governed by MLS IDX policies. Just as a visitor would enter a real estate office and provide "consumer information" to learn about local listings, so does VOW. Site visitors on an agent/agency website, that uses the VOW business, must enter in some form of contact information before accessing MLS IDX listings. That model turns the site visitor into a "customer", and acting as a lead-generation resource.

ILD to Replace VOW and IDX

Under the Internet Listing Display (ILD) policy enacted for July 1, 2006, listing brokers will not be allowed to restrict the display of their listings on selected competitors' Web sites, as under the provision called "selective opt-out" contained in NAR's Virtual Office Website (VOW) policy. VOW has been superceded by the ILD policy.

> "...*Known as Internet Listing Display (ILD), the new policy consolidates and replaces both the VOW policy and NAR's Internet Data Exchange (IDX)*

---

[67] Field Guide to Virtual Offices Web sites (VOWs) and Internet Data Exchange (IDX). National Association of Realtors. http://www.realtor.org/libweb.nsf/pages/fg902 (Retrieved 25 Apr 2005).

[68] NAR's Virtual Office Web Site Education Center. National Association of Realtors. http://www.realtor.org/realtororg.nsf/pages/vowhome (Retrieved 10 Dec 2005).

policy to create a single, unified policy governing the Internet display of all property information originating from the more than 800 multiple listing services owned and operated by Realtor® organizations. All Realtor® MLSs will be required to comply with the new policy by July 1, 2006...[69]

The Internet Listing Display (ILD) looks to establish equity and fairness in the display of MLS Listings on approved websites. Brokers are not allowed to "opt-out" from their competitors displaying their listings on the competitor's website. There's much to come after July 1, 2006 in terms of how technology providers handle this policy change that MLS-providers, brokers, and agents must adopt.

---

[69] New NAR Policy Delivers More MLS Information to Consumers. Steve Cook. National Association of Realtors. http://www.realtor.org/PublicAffairsWeb.nsf/Pages/NewMLS Policy?OpenDocument (Retrieved 28 Dec 2005).

Chapter

10

Neighborhood Search

Neighborhood search, incorporated into an agent's website, helps to establish a full service website for home buyers and sellers. Not only is the buyer's search for the perfect home imperative, but so is the search for the perfect neighborhood. Value-added resources such as neighborhood search found inclusive of an agent's website, keep the visitor on the website longer. Also, such a resource mitigates electronic communication between buyers & sellers and the target agent, providing a "lead generation" tool within the agent's website.

Commonly, real estate consumers would have to pay for such market knowledge from neighborhood search websites. Rather, an agent establishes a subscription and then incorporates the neighborhood search from their website. Here, the agent "foots the bill" for a specified number of searches for a given month, all-the-while obtaining leads sent directly to the sponsoring agent.

Once placed on an agent's or broker's website, no additional maintenance is required on behalf of the agent. Agents respond to client requests (i.e. sent to the agent's email inbox) to see homes in their chosen neighborhoods or communities. A solid provider of neighborhood search, added to an agent's website, does not charge any referral fees or lead generation fees. Thus, subscribing real estate professionals can save thousands of dollars when compared to the large referral fees from real estate, lead websites.

Lead-Generation

Sixty-eight percent (of agents & brokers) indicated they receive Internet leads, yet seventy-two percent are not satisfied with the amount or quality of the Internet leads they receive.[70]

Let's examine two overarching website-types for lead generation: (1) consumer websites and (2) an agent's website. Many consumer websites provide agents or brokers lead-generation "networks", whereas lead-generation techniques on an agent's website are directed to that agent. Both techniques can be fruitful in obtaining leads from interested buyers and sellers.

Providing access to neighborhood search, in both scenarios, has become a popular and must-have trend amongst agents looking to obtain leads. Let's first examine lead-generation on consumer websites, before showcasing how neighborhood search play's a role in an agent's website.

Comsumer Websites

Consumer websites that provide leads to real estate agent professionals are websites that appeal to the "curious" consumer. Site visitors may access real estate market information and even property valuations. Commonly, the consumer must commit to an agent contacting them for the consumer to access this market information. There is the perception that consumers may visit these websites more than MLS or broker websites, simply due to their "neutrality" in acting as a source of real estate information.

The issue that has arisen from these consumer websites, those sites that provide leads for agents and brokers, is sometimes their lack of quality leads. This is partly due to due "minimal" screening of these leads before reaching an agent or broker in the network (there are some consumer websites who *do* provide quality leads to agents and brokers, such as Homegain). Although these sites appeal to consumers doing market research, sometimes, that's all the consumer desires *is* market research. On rare occasions, some consumers will "agree" to guidelines of the lead-generation websites allowing an agent to contact them, when they are not really interested in an agent doing so.

---

[70] *2005 REALTOR® Technology Efficiency Survey.* Center for REALTOR® Technology (CRT). Published May 03, 2005, p34. http://www.Realtor.org/CRT.

A best practice for agents and brokers in reviewing lead-generation (consumer-oriented) websites is to inquire about their screening process of leads.

Resources!

NeighborhoodScout.com has built a comprehensive network of brokers called "NeighborhoodScout Broker Network" who service referrals that originate from the NeighborhoodScout.com website. Leads come from consumers looking to find homes in ideal neighborhoods, neighborhoods they've found directly at NeighborhoodScout.com.

Those leads are reviewed, screened, and "polished" before being referred to the real estate professional by participating brokers. Such care in providing quality referrals is not always found with lead-generation (consumer-oriented) websites

---

As described in Chapter 5: *Agent Website (Main Presence)*, agent websites provided in-part by *Webographer-friendly hosting companies*, come with an array of lead generation tools. With the case-study of Rapidlistings.com, agent websites may natively include, "Leads 2 Cell", "Leads 2 Email", "Contact Forms", etc.

As detailed in Chapter 6: *Single-property Websites*, contact forms are "standard issue" with every property website. Throughout single-property websites, are inclusive web pages that "spark" electronic communication between the agent and potential buyer.

Chapter 6 detailed how to generate "traffic" to your agent website. However, traffic and more site visitors are only as good as the leads generated from that traffic. Are there additional resources that can be added to an agent's website or single-property websites to further provide leads from such web site traffic? Let's examine how providing neighborhood search is a way to maximize the traffic to your agent website or single-property websites and in-turn, generate leads.

Neighborhood Search

Imagine having a neighborhood search engine on your agent website, for potential buyers and sellers that covers all 61,000 neighborhoods in America. Where search results provide all-inclusive community information such as: detailed street maps, median house values, neighborhood appreciation rates, exclusive school district ratings, FBI crime rates, and more, for every neighborhood in America.

Imagine this neighborhood search on an agent's website converting web traffic into leads that are delivered right to the agent/assistants email (or drip marketing program), with inquiries on additional community information or homes in a selected neighborhood. Simply stated, NeighborhoodScout.com provides this exact neighborhood search service and has made it available for agents and brokers to integrate into their websites.

A value-added resource like neighborhood search that can be integrated into an agent's website is made available in stellar fashion by Location Inc. via their flagship application called NeighborhoodScout, made available exclusively at www.NeighborhoodScout.com. NeighborhoodScout currently powers the Neighborhood Search of major consumer websites, such as HomeGain.com and ZipRealty.com. That same neighborhood search engine, used by industry giants, is now made available to agents and their respective websites.

Having neighborhood search on an agent's website, ignites electronic communication between buyers & sellers with the target agent; providing a "lead generation" tool within the agent's website. It's a topic of conversation that allows the potential buyer to "reach out" to the agent of their own free will. Such conversation by potential buyers can turn the target agent into the buyer's agent.

Showcasing your listed properties is imperative on an agent's website. However, the search for the perfect neighborhood has equal value to today's buyer. Again, the search for the perfect home is important; however, the search for the perfect neighborhood is equally imperative to today's web-savvy buyer

The rest of this chapter takes you through an interactive case-study of NeighborhoodScout and its powerful neighborhood search engine. This is provided as a powerful case-study to demonstrate what it means to add such a resource to your agent or agency website and to showcase its breadth and depth of neighborhood and lifestyle information made available.

Case-study of NeighborhoodScout

Let's consider a site-visitor looking for a luxurious community and lifestyle in Arizona. The example in Figure 31 on the next page, considers a search of with a 10-mile radius of Scottsdale, Arizona; known for its well-to-do citizens, lavish lifestyles and luxurious homes.

Neighborhoodscout.com search is simple to use from the perspective of site users. After selecting the overarching lifestyle search (i.e 'Luxury Communities"), Figure 31 solicits basic search criteria for a given city and state. Let's now examine the neighborhood results for Scottsdale, Arizona.

The search results, as displayed in Figure 32 on the next page, produced neighborhoods that met "luxurious" criteria: Median House Value of $500K-$750K, with a population that is Educated, Wealthy, and are Executive, Management, & Professional-types. The map with darker shaded areas represents a "closer" neighborhood match than the lighter shaded areas. "Paradise Valley", Neighborhood #2 is selected in that case-study example.

Figure 31: Luxury Community Search—NeighborhoodScout.com [71]

---

[71] *Luxury Communities*. Location Inc. Retrieved December 10, 2005. Reprinted with permission. http://real-estate.neighborhoodscout.com/option_3.jsp.

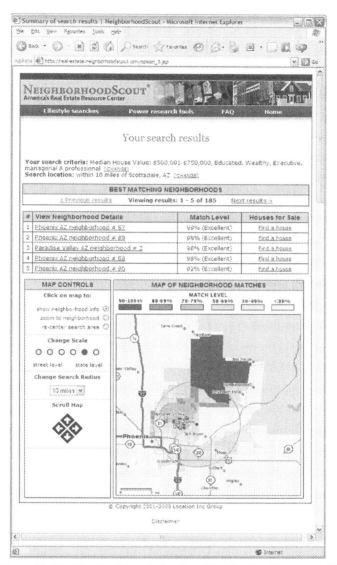

Figure 32: Neighborhood Search Results—NeighborhoodScout.com [72]

[72] *Your Search Results.* Location Inc. Retrieved December 10, 2005. Reprinted with permission. http://real-estate.neighborhoodscout.com/option_3.jsp.

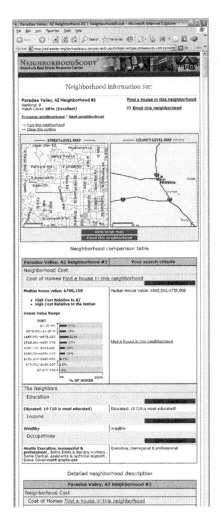

Figure 33: Specific Neighborhood Search Results—NeighborhoodScout.com[73]

---

[73] *Neighborhood Information for Scottsdale, Arizona.* Location Inc. Retrieved December 10, 2005. Reprinted with permission. http://real-estate.neighborhoodscout.com/servlet/OutputInfoServlet

Figure 33 displays the depth of neighborhood and community data presented to a user of the NeighborhoodScout search engine. At the top of the Neighborhood Information page for Scottsdale, Arizona, there is a hyperlink for "Find a Home in this Neighborhood". This hyperlink takes the site visitor to a (1) contact form provided by NeighborhoodScout or (2) the agent's own drip marketing program.

Lead-Generation—Agent's Web Presence

An agent who adds NeighborhoodScout to their website, receives leads from a "personalized" contact form (or their hyperlinked drip marketing program), when a user selects *"Find a Home in this Neighborhood"*. Let's begin to analyze overarching techniques of how to add neighborhood search to an agent website.

Neighborhood Search on an Agent's Website

After sign-up, NeighborhoodScout sends a welcome message, providing information on how to set-up NeighborhoodScout on your agent website in addition to accessing their support website. Adding NeighborhoodScout can be as simple as adding a hyperlink on an agent's website that redirects users to a specific web page hosted at NeighborhoodScout.com. The core URL that is important to adding NeighborhoodScout to an agent's website includes:

> http://real-estate.neighborhoodscout.com/servlet/LoginDBPseudoServlet?
> username=**xxxxxxxxx**&password=**yyyyyyyyy**

However, the full URL that you must use contains your "public" username and password, as found in the welcome email to new subscribers. An example of this includes:

> http://real-estate.neighborhoodscout.com/servlet/LoginDBPseudoServlet?
> username=714294559&password=**abcdwxyz**

This URL loads a personalized, neighborhood search experience for an agent's website, and user-base. The "public" username and password, included in the URL, means this search page is provided in-part by the agent who subscribed to the service. All "leads" by site visitors that are generated from the associated URL, go to the email address (or drip marketing program) associated with that agent/broker subscriber.

Once made available on an agent's or broker's website, no additional maintenance is required on behalf of the subscriber. Agents respond to client requests (sent to the agent's email inbox or responsible person of the drip marketing program) for their chosen neighborhoods or communities. A solid provider of neighborhood search, added to an agent's website, does not charge any referral fees or lead generation fees. Thus, subscribing real estate professionals can save thousands of dollars when compared to the high referral fees on real estate leads.[74]

Does the setup sound complicated? Well have no concern. Neighborhood Scout.com's *Agent Subscription Support Site* for subscribers, has just what you need to add NeighborhoodScout to your website in seconds!

Neighborhood Search—Link Everywhere

Given that a NeighborhoodScout agent subscription is only $2 per day, an agent should make a point to reference their personalized URL in every possible outlet of their web presence. Such outlets in an agent's web presence where personalized, neighborhood search should be made accessible includes: the agent website and on each single property website.

Agent Website

A RapidListings.com agent website can support framing of any external web page. Figure 34 shows how the NeighborhoodScout search engine is incorporated within the agent's website for a seamless appearance.

---

[74] *Real Estate Agents and Brokers: Add NeighborhoodScout to your Website Today.*. Location, Inc. Retrieved December, 10 2005. http://www.neighborhoodscout.com/real-estate-leads.jsp.

Figure 34: NeighborhoodScout Inclusive of RapidListings.com, Agent Website

Single-property Websites

With any AgencyLogic.com PowerSite (i.e. www.123AnySt.com), custom "menu" buttons can be added to the navigational menu. The buttons can take a user to a web page either you created, or act as a hyperlink to an external web page. The latter technique can be used to make an agent's NeighborhoodScout subscription accessible to site visitors, as shown in Figure 35.

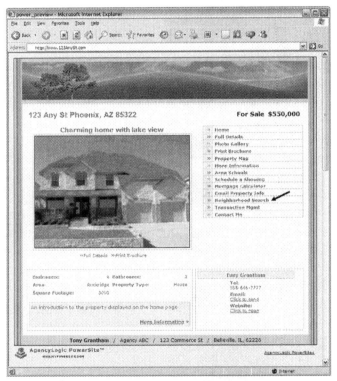

Figure 35: An AgencyLogic PowerSite including
a NeighborhoodScout link on the navigational menu

NeighborhoodScout.com

NeighborhoodScout.com makes it easy for real estate professionals to turn their websites into a full service real estate destination for home buyers and sellers by unveiling a new subscription that integrates NeighborhoodScout® into their own websites. Lead-generation is paramount to an agent building their own web presence. NeighborhoodScout helps agent subscribers generate *exclusive* real estate leads from their own website traffic. The search engine covers all 61,000 neighborhoods in America. NeighborhoodScout allows agents to provide detailed street maps, median house values, neighborhood appreciation rates, exclusive school district ratings, FBI crime rates, and more, for every neighborhood in America, with one single application that goes on the agent's website.[75]

Activity!

At Webographers.com, candidates of the REAL ESTATE WEBOGR APHER™ receive their own NeighborhoodScout®, demo subscription. Candidates examine the ease in setup of NeighborhoodScout on their demo agent website, and demonstrate its lead-generation capability with a practice "dialogue" with a course proctor.

One of a Kind

There simply is no other dynamic, neighborhood search resource that can be added to an agent or broker's website like Neighborhoodscout®. Why is this web-based, neighborhood search of NeighborhoodScout one-of-a-kind? This all inclusive neighborhood and community search engine is *patented technology*, designed to allow individuals and families to instantly find the neighborhoods that are best for them in any area of United States they choose.

Intelligent Technology

Traditionally, agents may add neighborhood and community information on a web page they design; information manually typed by the agent. This information is "static", rarely updated, and doesn't natively solicit leads from site visitors. For some site visitors, they may see it as "fluff" incorporated by the agent to add another web page to an already bare-boned website.

NeighborhoodScout® is a *dynamic* search engine that employs a unique way of matching people with the best neighborhoods for them and their families. What do we mean by *dynamic*? The search page, added to agent websites is powered by the largest database of neighborhood characteristics ever created with impeccably accurate matches.

---

[75] *Real Estate Agents and Brokers: Add NeighborhoodScout to your website today.*. Location, Inc. Retrieved December, 10 2005. http://www.neighborhoodscout.com/real-estate-leads.jsp. Reprinted with Permission.

The best matching neighborhoods are instantly mapped for convenience to the user. Every element of Neighborhood data returned in the search results are matched against the users search criteria. On the generated map, areas closely matching consumer's criteria are shaded "darker" versus the "lighter" shaded regions of lesser matching neighborhoods. This enhanced mapping generated in real time, was displayed in Figure 32 found earlier in this chapter.

Rather than displaying raw data, NeighborhoodScout employs a user-friendly, graphically enhanced interface. Adding Neighborhoodscout to your agent or broker website shows you put your clients and site-visitors interests first. Not only do you assist in clients finding the perfect home, but you ensure they also find that perfect neighborhood.

The Data

The data Location Inc® uses for NeighborhoodScout® is the latest available from several leading government sources. These sources include the U.S. Bureau of the Census, the U.S. Department of Justice, the National Center for Education Statistics, and the U.S. Geological Service, among others. This means that data is current and up-to-date for the site visitors who utilize NeighborhoodScout made accessible from an agent or broker's website.

NeighborhoodScout® uses census tracts as their designated neighborhood unit. Since census tracts are *sub zip code*, they allow for better neighborhood matching; helping consumers find that ideal community.

> *"Census tracts are small, relatively permanent subdivisions of a county that are defined by the U.S. Census Bureau. Census tracts usually have between 2,500 and 8,000 persons and are defined to contain areas with homogeneous population characteristics, including economic status, and living conditions."*[76]

NeighborhoodScout® uses nearly 200 characteristics to describe each and every neighborhood (census tract) in America. These include:

- school quality
- housing costs
- crime rates

---

[76] *About our Neighborhood Search Engine.* Location, Inc. Retrieved December, 10 2005. http://www.neighborhoodscout.com/neighborhood-search.jsp. Reprinted with Permission.

- income levels
- the age, size and style of homes
- the density of buildings
- rental areas versus owner occupied
- the proportion of families with children
- the ages of persons in the neighborhood
- ethnic and racial makeup
- educational levels
- languages spoken
- types of careers of those living in the neighborhood
- numbers of farms
- many more

This set of parameters truly encompasses the "character" of neighborhood. Today's real estate consumer is not just concerned with finding that 4 bedroom home with 2 baths. The consumer can be choosy in the neighborhood that a potential property resides. Neighborhood search as employed on an agent or broker's website and powered with NeighborhoodScout, appeals to such consumer interests.

NeighborhoodScout: History & Evolution
NeighborhoodScout® was created by Dr. Andrew Schiller, who earned his PhD from Clark University's Graduate School of Geography, *America's oldest and largest geography PhD program*. Dr. Schiller was previously a scientist at Oak Ridge National Laboratory and Atomic Energy Complex, and also Director of Science for The Nature Conservancy's Tennessee Chapter.[77]

A team of experts in computer mapping & analysis and web design joined Dr. Schiller at NeighborhoodScout. The coupling of Information Sciences, User Interfaces, and Geography has created an "explosive" neighborhood search engine; establishing the most popular resource any agent or broker can add to their website. Agents and brokers who add NeighborhoodScout to their website will have their consumers *thanking* them for putting their best interests first!

---

[77] *About our Neighborhood Search Engine.* Location, Inc. Retrieved December, 10 2005. http://www.neighborhoodscout.com/neighborhood-search.jsp. Reprinted with Permission.

Sales Information

To establish an agent or broker account subscription and add NeighborhoodScout® to your website, visit http://www.neighborhoodscout.com/real-estate-leads.jsp. Also visit https://www1.neighborhoodscout.com/ssl/realtors/ to join their "Broker Network" to start receiving referrals from site-visitors of NeighborhoodScout.com. For immediate sales inquiries, feel free to also contact via email at info@NeighborhoodScout.com.

Chapter 11

CMA's and AVM's

Comparable Market Analysis (CMA) is a report generated by agents and brokers that include: (1) a market value of a property and (2) comparable sales information of properties with similar "size" and "shape". The market value is an educated estimate by an agent, guided by an agent's own personal knowledge of market activity and the use of recent comparable sales.

Many agents refer to the CMA in two formats, the *Seller's CMA* and the *Buyer's CMA*. The Seller's CMA has a short "shelf-live", utilized to win-the-listing by showing preparedness in market knowledge and property valuation to a potential seller. Buyer's CMA includes the Seller's CMA information, but expands by including "attractive information" like local schools, major businesses, area attractions, extensive photos of the target property, and more. Buyer's CMA will likely have a longer shelf-life, essentially until the property is sold.

Commonly, the MLS is used to extract a CMA, but many agents feel this document is not "print-ready" to give to consumers. Thus agents and brokers will develop their own CMA, using the MLS data and other repositories like county records, tax records, etc to develop a custom report. From this data, agents and brokers will establish a property value based on gut-feeling, commonly referred as a *guesstimate*. What are some of the efforts that agents may go through to develop a custom CMA? Let's take a look at a real life example.

CMA Development Process

Nancy E. Chadwick, President and Broker of Chadwick Real Estate, Inc., is a PA licensed real estate Broker and Instructor, has her own process. She calls her process "The Art of the Residential CMA". A paraphrased summary of her process includes 6 key areas: Location, Time, Housing Style & Use, Size, Rooms, and Conditions & Amenities.

1. Location—the comparables should closely match the subject's location. If it's necessary to extend the search radius, the search should not go into other municipalities and school districts.

2. Time—Nancy tries to use comparable sales data less than 6 months old. Any older, she states will require adjustments for older sales. In periods of a more steady market, she'll reach back further than 6 months to pull comparable sales.

3. Housing Style & Use—She remains with comps that are the same style as the target property in discussion (same number of stories, ranch, contemporary, duplex, split level, etc).

4. Size—both the house and the lot are equally important in identifying comparable sales. For differences in house square footage (at or above grade areas only), she uses a factor of a given dollar amount per sq footage (i.e. $35/sq ft). The same is true for the lot size, where in Pennsylvania; she'll use a factor of $10K per acre.

5. Rooms—She'll adjust for differences in the number of bedrooms, number of baths, family room, living room, office, garage size, central air, gas vs. electric, utilities and basement/attic

6. Conditions & Amenities—Adjusting for these elements is complicated because her evaluation of comps is usually based on: drive-by evaluations, supplemental information from the MLS, and database information and conversations with the agents involved in the transaction.

Finally, Nancy's CMA utilizes an Excel spreadsheet, containing dollar adjustments for each property characteristic (*such as age, lot size, house SF, basement, number of bedrooms, number of bathrooms, garage size, location, utilities, date of sale, amenities, and more*).[78]

---

[78] *The Art of the Residential CMA. Nancy Chadwick.*. Nancy Chadwick. Obtained December, 10 Copyright 2002, 2003. http://www.reiclub.com/articles/residential-cma.

Does your CMA process follow similar to Nancy? Is it different? As one may conclude, "each agent has their own 'special sauce' when doing listing presentations with customers."[79] Many agents look to create a CMA document inclusive of their personal/corporate branding, property valuation & comparable properties, graphs & charts, and sometimes inclusive of neighborhood information: such as school, business, and community. All-in-all, the process can be time-consuming and tedious.

Issues with the CMA from the MLS

Many issues exist with the CMA-creation process as highlighted by the extensive research from the Center of Realtor® Technology (CRT). The two focal reports called *2005 REALTOR® Technology Efficiency Study* dated February 23, 2005 and May 03, 2005 respectively, give insight to the issues related to today's CMA. Many claims are made in the next sub-sections, and are supported by citations from these two reports.

Data Accuracy

Typically, data found in the MLS can be skewed, error-prone, inaccurate, or incomplete. Due to such issues, brokers, agents, and support staff must make adjustments to the comparables provided. Or they must "dig" and "comb" through the MLS manually to find comparable sales they desire.

> "…As an example, private sales listed for $1 (due to owners not being willing to list a sale price) throw off the CMA statistics yet they cannot eliminate these listings from the calculations…"

> "…For example, MLS systems do not account for homes listed with no sales price. This throws off the calculation of listing or sales price per square feet. When agents prepare CMA's they add in the real sales price or make necessary adjustments that are not possible with the MLS system…"[80]

---

[79] 2005 REALTOR Technology Efficiency Study. (2004, February 23), WAV Group in conjunction with Center for REALTOR® Technology (CRT). p 17. http://www.Realtor.org/CRT

[80] 2005 REALTOR Technology Efficiency Study. (February 23, 2004). Center for REALTOR® Technology (CRT). p6 and p19 respectively. http://www.Realtor.org/CRT

Need to Supplement the MLS

Given the MLS and its standard CMA-tool, non-MLS data may be required as a supplemental resource. In addition, the included reporting tools are not conducive to "making it your own", where agents buy 3rd party applications to generate a more "print-ready", CMA report.

> *"...14% suggested improvements to comparable analyses. This group also requested access to non-MLS data like tax information and adjustments for private sales...."*

> *"...49% of respondents use other tools to extend the MLS supplied CMA..."*[81]

Lack of Flexibility

Many are going outside the MLS to obtain comparable sales data known to be accurate, like county records and tax information. In addition, lack of customizing the MLS-supplied CMA, has brought agents to use additional 3rd party tools and data sources.

> *"...Agents interviewed believe CMA (Comparative Market Analysis) and statistical tools could be more effective. A lack of flexible reporting tools creates a need for agents to manually produce analyses of market and neighborhood data..."*

> *"...Nearly every agent in the study said the MLS CMA was not flexible enough to allow for a variety of custom analyses needed to properly present the correct market picture. They would like to look at statistics like average days on market, asking vs. selling price average, dollars per square foot and have more ability to compensate for listing anomalies..."*[82]

Lack of Personalization

MLS provided CMA tools don't normally allow for the incorporation of personal or corporate branding. Such capabilities would allow the agent or broker to

---

[81] 2005 REALTOR Technology Efficiency Study. (May 03, 2005), Center for REALTOR® Technology (CRT). p 31 respectively. http://www.Realtor.org/CRT

[82] 2005 REALTOR Technology Efficiency Study. (February 23, 2004). Center for REALTOR® Technology (CRT). p4 and p6 respectively. http://www.Realtor.org/CRT

"make it their" own. Such personalization may include a mission statement, photo, corporate logo, contact information, etc.

> *"...Most listing presentations [CMAs] included company information, agent biography, a marketing plan and some market information..."*[83]

> *".. The need for flexibility and personalization of CMAs is sited as the reason many are using third-party or self-developed methods and tools..."*

> *"...Respondents also believe that their unique personalized CMA presentations help differentiate them from their competition..."*

> *"...The ability to personalize the CMA is the number one improvement suggested. The next requirement is the ability to make adjustments, followed by ease of use. CMA personalization may be why users build their own CMAs or use third party products...."*[84]

What the above research findings show is that the MLS CMA does not commonly produce definitive CMA's that are print-ready for distribution inclusive of agent "personalization". Rather, they are a source of data and information, where the data is "cut & pasted" and manipulated in a word processing document, for example.

Deficiencies in Data Representation

Commonly, MLS CMA's don't include data representation tools such as charts and graphs. As with the example of Nancy Chadwick in the beginning of this chapter, she will commonly work with Microsoft Excel to produce a CMA. With data displayed in a spreadsheet application, charting and graphs can be manipulated by the common user. When using a word processor for your print-ready CMA, such graphs and charts will have to be cut & pasted from the spreadsheet, into the word processing document.

> *"...Often, they will prepare detailed "trend analysis"* [as supported by graphs] *on their own to show how areas of the city are changing from year to year to help justify their pricing and marketing recommendations. The data*

---

[83] 2005 REALTOR Technology Efficiency Study. (February 23, 2004). Center for REALTOR® Technology (CRT). p17 respectively. http://www.Realtor.org/CRT

[84] 2005 REALTOR Technology Efficiency Study. (May 03, 2005), Center for REALTOR® Technology (CRT). p4, 6, 32 respectively. http://www.Realtor.org/CRT

is also not easy to export into a spreadsheet which can be customized by the agent or assistant... "[85]

30% of respondents believe "supporting graphics" is the number 1 desired improvement to MLS CMA-tool. [86]

Conclusion

Given all the issues noted as of recent with the MLS supplied CMA-tool; there are many areas to improve the process of CMA generation. Some of the issue can be improved with technology. Furthermore, to note how new technology can improve the CMA-process; one must examine the business use and intention of the CMA, i.e. Seller's CMA vs. Buyer's CMA.

Market Solutions for the CMA

As stated in Chapter 9, MLS repositories must adhere to the Real Estate Technology Standard (RETS). This enables 3rd party applications to work "with" the MLS, commonly to extract property data to another application.

Given that scenario, there are applications that sit "on top" of the MLS to generate a more "print-ready" CMA than that of the MLS-provided CMA. Such applications allow agents to provide personal and corporate branding, and sometimes data representation such as graphs and charts. What such systems cannot inherently do is *correct* the anomalies found with MLS data. As previously stated, MLS data may be error-prone from human data-entry that requires adjustments by those creating a CMA.

The next evolution in CMA-generation process comes from the use of the Automated-Valuation Model (AVM) report. Although numerous methods exist to generate a print-ready CMA or to simply find supplemental data, the remainder of this chapter and in-depth case-study focuses on this technology.

---

[85] 2005 REALTOR Technology Efficiency Study. (February 23, 2004). Center for REALTOR® Technology (CRT). p19. http://www.Realtor.org/CRT

[86] 2005 REALTOR Technology Efficiency Study. (May 03, 2005), Center for REALTOR® Technology (CRT). p32. http://www.Realtor.org/CRT

Automated Valuation Model (AVM)

Automated Valuation Model is a process to establish a market value of a property through "scientific" measures, versus a human's inclination. AVM reports will commonly include property valuation, comparable sales, and demographic information. Data sources of AVM providers may include public record information databases of county, tax records, census, in addition to other proprietary databases.

An automated valuation is not an appraisal and does not include an inspection in the analysis. Rather, AVMs commonly use public record information sources, and rigorous scientific algorithms to produce property valuations. AVMs are commonly used by the financial and lending industry for loan evaluation.

AVM vs. CMA

In short, the AVM and CMA (personalized by the agent using the standard MLS-CMA tool) have many differences and similarities. The table below looks to engage those comparisons, and provide details into better explaining the AVM, which may not be of readily understood by real estate professionals.

AVM vs CMA		
Qualitative Comparison Chart		
Item	AVM	CMA
Data Source	Public information records (i.e. county, tax records, proprietary databases)	MLS *(commonly)*
Property Valuation Model	Generated by scientific algorithms and formulas	An estimate based on the agent's market knowledge
Primary Audience	Lending & financial institutions, consumers, real estate professionals	Consumers, real estate professionals
Property Search Criteria	Mailing Address	Sq footage, # of bedrooms, lot size, # of bathrooms, amenities, garage size, property type, avail. rooms
Time-to-Generate	Seconds	Speed of the real estate professional *(sometimes lengthy if a custom report)*
Accuracy	Highly Accurate	Dependent upon the accuracy of data and the search techniques in MLS
Current Comparable Sales Data	Commonly a 30-day delay before sale information is registered into databases	Real-time from MLS
Trend Analysis	Sometimes included	Manually performed
Charts & Graphs	Sometimes included	Manually performed
Agent Personalization	Never before included *(until now)*	Manually performed

AVMs in the Lending Industry

AVMs are a popular resource in the financial and lending community. They are utilized to validate and supplement the findings of traditional appraisals typically used in the underwriting process. Certain types of AVM products can be used in lieu of traditional appraisals when financing requires an expedited turn-around. Traditional appraisals cannot be fully replaced by AVM technology from the eyes of the lending community. Rather, AVMs are used to supplement appraisal information and to serve as an additional research tool.

AVMs in the Consumer Market

Many consumers are Internet savvy and eager to perform their own market research, from both the seller and buyer perspectives. AVM data has become widely adopted by consumers when conducting research to determine the value of a given property.

For households looking to put their homes on the market, they may conduct initial research to see "what their home is worth". For potential buyers, AVM data may give them insight about the lending range for a specific property. Many websites providing "home valuation" services to consumers commonly include:

1. *Lead-generation websites*—Although they are not referred to as "lead-generation websites" that is exactly what they are. They offer free valuations to consumers, in exchange for the ability of an agent in their network to solicit the consumer with real estate services. Many times, these consumer websites utilize AVM data to generate property valuations and comparable sales.

2. *FSBO listing websites*—to provide value-added resources that support for-sale-by-owners, many FSBO websites will include an AVM report to help sellers define their "listing" price. Sometimes, this is included in the FSBO listing package or it may be offered as a supplemental service. For buyers perusing the FSBO listings, purchasing a home valuation report allows them to validate the FSBO listing price.

3. *Bank & lending websites*—to entice potential buyers to apply for home loans online, many online banking and lending institutions may provide basic property value reports and comparable sales information. This is done to engage potential buyers in need of financing and to simply provide a value-added, user-friendly feature on their website.

4. *Consumer AVM websites*—Consumer AVM websites provide property information to potential sellers considering putting their home on the market. Conversely, the report allows buyers to cross-check a listing price for a desired property. The type of property information found in these types of reports is shown in Figure 36 on the next page. There is usually a nominal charge for a full-report ($19 to $30). The consumer can conduct research at their convenience and on their own time.

The depth of information displayed in Figure 36 on the next page shows why the AVM has been widely adopted by consumers and financial institutions alike. The next logical evolution is the use of AVM technology by real estate agents, brokers and their support staff.

How can an AVM's improve efficiency in the daily activities of the real estate professional? When viewing the consumer AVM in Figure 36, an agent or broker may conclude that it "looks", "smells" and "tastes" very much like a typical CMA. Property value and neighborhood information are included in the AVM. In addition, comparable sales, maps and other value-added information will generally be included.

It's these observations that continue to increase the use of AVM technology by agents, brokers, and assistants in the CMA process.

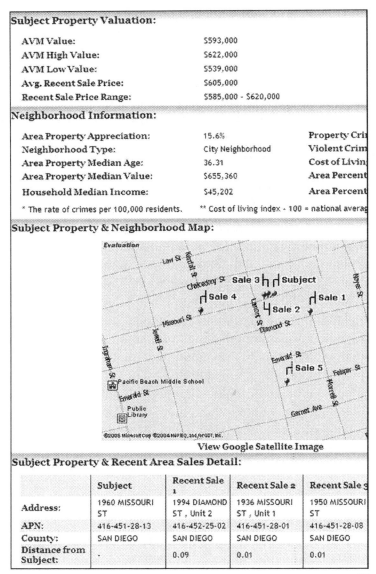

Subject Property Valuation:

AVM Value:	$593,000
AVM High Value:	$622,000
AVM Low Value:	$539,000
Avg. Recent Sale Price:	$605,000
Recent Sale Price Range:	$585,000 - $620,000

Neighborhood Information:

Area Property Appreciation:	15.6%	Property Cri
Neighborhood Type:	City Neighborhood	Violent Crim
Area Property Median Age:	36.31	Cost of Livin
Area Property Median Value:	$655,360	Area Percent
Household Median Income:	$45,202	Area Percent

* The rate of crimes per 100,000 residents.   ** Cost of living index - 100 = national averag

Subject Property & Neighborhood Map:

View Google Satellite Image

Subject Property & Recent Area Sales Detail:

	Subject	Recent Sale 1	Recent Sale 2	Recent Sale 3
Address:	1960 MISSOURI ST	1994 DIAMOND ST , Unit 2	1936 MISSOURI ST , Unit 1	1950 MISSOURI ST
APN:	416-451-28-13	416-452-25-02	416-451-28-01	416-451-28-08
County:	SAN DIEGO	SAN DIEGO	SAN DIEGO	SAN DIEGO
Distance from Subject:	.	0.09	0.01	0.01

Figure 36: Comsumer AVM—Devenio.com, *a service of eAppraiseIt LLC* [87]

[87] *Consumer AVM Report*. Devenio™, a service of eAppraiseIt, LLC. Retrieved January 08, 2005. Reprinted with permission. http://www.Devenio.com/sample_report.htm.

AVM Use by Agents & Brokers

Extensive research conducted by the National Institute of Webographers, LLC concluded that the next logical evolution in the use of AVM technology would be among real estate agents, brokers and their assistants as part of the CMA-process. In the early part of this chapter, the section labeled *"Issues with CMA from the MLS"* highlights deficiencies, problems, and manual "work-arounds" when using the MLS CMA-tool.

The MLS CMA-tool is a good resource for conducting basic research for a local market. Comparable sales are used to develop a presentable CMA to buyers and sellers. However, it has been noted that deficiencies exist in the categories of: 'Data Accuracy', 'Need to Supplement the MLS', 'Lack of Flexibility', 'Lack of Personalization', and 'Deficiencies in Data Representation (*charts & graphs*)'.

Can the AVM replace the MLS CMA-tool entirely? The answer to that question is generally no. However, AVMs can augment a custom CMA by providing supplemental comparable sales data and by providing a value range the agent can use to validate their existing estimation of the property value. Depending on the activities of the real estate professional, the AVM can be generated instantly and serve in-lieu of a CMA when time does not allow for the generation of a traditional CMA.

Provider—AVMs for Agents & Brokers

The National Institute of Webographers had established a vision for a new service that utilizes AVM technology to assist agents in the CMA-process; a service mindful of MLS CMA deficiencies and other research as cited in this chapter. With a vision based on sound qualitative research in-hand, National Institute of Webographers desired an AVM provider who could to take the conceptual vision, design and requirements to produce a new service for real estate professionals never before seen on the market.

The National Institute of Webographers, LLC conducted an eight month review of AVM providers nationwide. This was done in an effort to determine who provides the best AVM services to consumers and the financial industry. The search also focused on specific competencies necessary to assist real estate professionals in generating CMA's; competencies that were in alignment with the National Institute of Webographers evolutionary vision.

eAppraiseIt, LLC

That search concluded with the selection of eAppraiseIt, LLC. They are a national leader in appraisal management services and are a subsidiary of First American Corporation. eAppraiseIt's new line of agent AVM's can be found at www.AgentAvm.com.

AgentAVM.com is a new and exclusive product for agents and brokers; based upon data competencies and best practices developed for lenders and financial institutions nationwide and guided by the conclusions brought forth by the National Institute of Webographers. This new AVM service was designed with the real estate professional in mind and with specific deference to the CMA issues and "head-aches" reported by the Center for Realtor® Technology (CRT).

The National Institute of Webographers takes great pride in the conception of AgentAvm.com, as it displays an instance of a technology provider producing services that are mindful of the real estate practitioner. Since this product is one-of-a-kind, with no other AVM company providing this type of service at the time of writing this book, let's discuss in full detail, www.AgentAvm.com from eAppraiseIt, LLC.

AgentAvm

AgentAvm is eAppraiseIt's flagship application for real estate professionals. It is where the "CMA meets the AVM" by incorporating the agent's personal branding (photo, mission statement and contact information). The agent version of the AVM report also includes important property information such as a value range, neighborhood information and recent sales data.

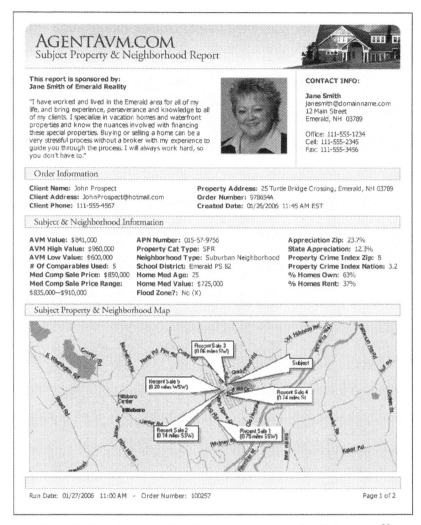

Figure 37: AgentAvm (page 1)—AgentAvm.com, *a service of eAppraiseIt* [88]

---

[88] *Sample Report.* AgentAvm™, a service of eAppraiseIt, LLC. Retrieved January 08, 2005. Reprinted with permission. http://www.AgentAvm.com/sample_report.htm.

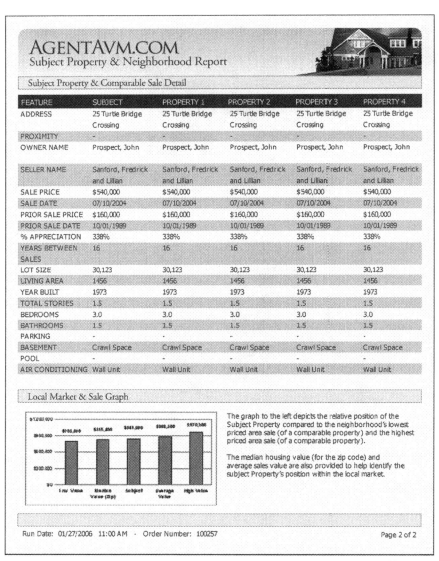

Figure 38: AgentAvm (page 2)—AgentAvm.com, *a service of eAppraiseIt* [89]

---

[89] *Sample Report.* AgentAvm™, a service of eAppraiseIt, LLC. Retrieved January 08, 2005. Reprinted with permission. http://www.AgentAvm.com/sample_report.htm.

Back-office Reporting

Agent accounts established at AgentAvm.com provide for back-office use in generating a full AVM report with personal branding in seconds, as shown in Figure 37 and Figure 38 on the two previous pages. This capability is extremely useful in generating a print-ready, CMA-*like* document when time is of the essence. For agents who create custom CMA documents, the report provides a supplemental source of valuation data and comparable sales.

Let's detail the advantages of AgentAvm.com for everyday, back-office use. These advantages are from the perspective of the Seller's agent and the Buyer's agent.

Seller's Agent

1. The AgentAvm can serve in lieu of a *full Seller's CMA,* especially when time is at a premium because the AVM can be generated in a matter of seconds. Considering that a Seller's CMA is designed to "win the listing" its lifespan is short-lived, perhaps no more than one day.

2. If the agent has a short "window" to meet a potential seller, ample time may not be available to generate a traditional CMA. AVMs only require the property's mailing address to generate a report, whereas a CMA requires substantially more information.

3. In "brick & mortar" agencies, office assistants do not need to solicit all types of information from potential seller's who walk-in or call by phone. All they need to obtain for the AVM is the property address. If an agent or office assistant is successful in obtaining specifications about a property using an MLS data sheet, the professional can "cross check" property specs, (i.e. number of bathrooms, sq footage and other data) that a potential client may have incorrectly stated. AVMs do report property specifications as well.

4. AgentAvm allows for agents to "cross-check" their own suggested listing value. Thus, the AgentAvm acts as a "second-opinion" to review the agent's suggested listing price. With inclusive "high" to "low" property values reported in the AgentAvm, the agent is now thinking in terms of what the prospective Buyer's purchasing range may be, since AVM data is widely used by lending institutions.

Buyer's Agent

1. AgentAvm can serve in lieu of a *full* Buyer's *CMA* if there is simply not enough time in an agent's schedule. Generating the report can "buy" the agent time until a custom, Buyer's CMA is completed.

2. AgentAvm can act as a supplemental source of market data and comparable sales information when completing a custom, Buyer's CMA. As stated earlier in this chapter, agents are using public information sources like county records and tax records to supplement and adjust the "holes" in MLS CMA reporting.

3. The Buyer's agent can use AgentAvm to cross-check the suggested listing price of other sellers. This second opinion helps to better evaluate if a property is ambitiously listed too high, appears to be a real value, or is lower than the reported value range as reported by AgentAvm.

4. Using the value range from the AgentAvm helps to evaluate the Buyer's possible purchasing range.

As a back-office tool, AgentAvm can replace and supplement many of the activities involved in generating a CMA.

Lead-Generation

Besides acting as a back-office tool, AgentAvm provides the agent with the ability to sell their personalized AVM's directly from their personal website. Similar to adding NeighborhoodScout.com's Neighborhood Search, the agent can truly make their website full service, and provide site visitors with home valuation services in real-time.

The capability of adding AgentAvm to an agent's website allows site visitors to continue to conduct their own market and property research. This is still a trend amongst consumers, wanting to reach out to agents when they are ready to do so. What's best is that results on valuation and comparable sales are delivered to the site visitor in real-time within a matter of seconds.

Although they obtain an AVM report from the agent's website, what's included in the report is the personal branding and contact information of the sponsoring agent. This further implies you are offering all inclusive services from your website, even home valuations.

Lead-Generation
In addition, the sponsoring agent receives an automated email message any time a consumer from the agent's website orders a report. AgentAvm from an Agent's website provides as an additional lead-generation tool. When the agent contacts this individual, the agent is not a random stranger, rather a person the consumer "knows", as they obtained a report from that agent's website. This helps the agent obtain a more "qualified" lead, likely to produce results.

AgentAvm—Link Everywhere

All in all, AgentAvm incorporated into an agent's web presence feeds the desire of consumers to "self-serve" their property and market research—while generating leads to the agent.

An agent should make a point to reference their personalized URL in every possible outlet of their web presence. Such outlets in an agent's web presence where the personalized AgentAvm webpage for consumers is made accessible includes: the agent website and single property websites.

Agent Website

A RapidListings.com, agent website can support framing of any external webpage. As displayed in Figure 39 on the next page, shows the AgentAvm webpage for consumers incorporated within the agent's website for a seamless appearance.

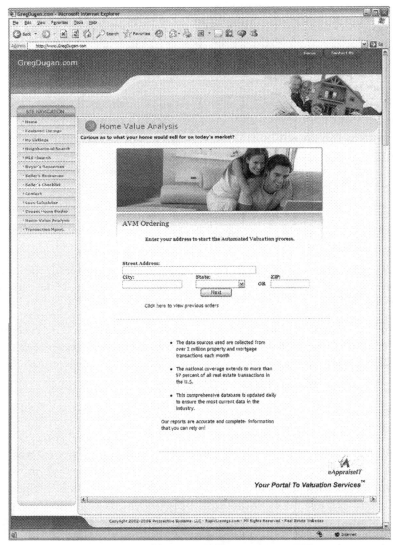

Figure 39: AgentAvm accessible in a RapidListings, Agent Website *(prototype)*

Single-property Websites

With any AgencyLogic.com PowerSite (i.e. www.123AnySt.com), custom "menu" buttons can be added to the website. The buttons can take a user to a webpage you created in the AgencyLogic control panel, or act as a hyperlink to an external web page. Just like the NeighborhoodScout example in the previous chapter, making the AgentAvm webpage for consumers accessible to site visitors of the single-property website, can be done as shown in Figure 40 below.

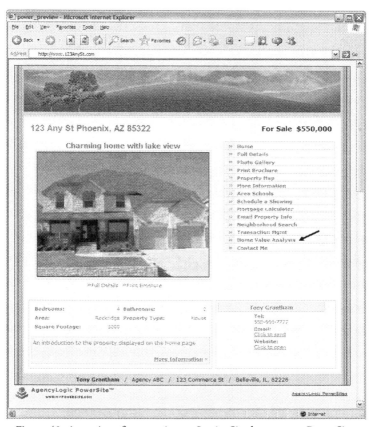

Figure 40: AgentAvm from an AgencyLogic, Single-property PowerSite.

AgentAvm.com

eAppraiseIt, LLC is the nation's leader in national appraisal management supporting the lending, appraisal and consumer industries. They have extensive experience with automated valuation products and they have access to the industries most comprehensive data resources for AVM generation. These automated valuation models are now being used by real estate professionals nationwide and can be ordered directly from their website at www.AgentAvm.com. This service provides AVM technology to real estate agents and real estate brokers. This new service utilizes the most advanced AVM technology available in the industry today. It is designed to add value to the valuation process for real estate professionals and lenders alike.

With agent or broker accounts, real estate professionals can generate a full "CMA-looking" report along with: agent photo, contact information, personal/corporate branding, comparable sales information, property value and demographic data in a matter of seconds. In addition, AgentAvm.com allows for agents and brokers to place a banner on their own website so visitors can order a report for a specific property. This type of "site" order also provides the added benefit of capturing consumer contact information that can be used for future solicitation purposes.

Activity!

At Webographers.com, candidates of the REAL ESTATE WEBOGR APHER™ certification work with AgentAvm.com. Candidates generate sample AgentAvm(s) to examine its value for use by the agent.

Supported by a Fortune 500 company, this California-based organization focuses on innovative electronic valuations and has been in the residential appraisal business for over 10 years.

PART FIVE:
The "Paperless" Transaction

Chapter 12: Electronic Forms

Chapter 13: Online Transaction Management

Electronic Forms

Electronic transactions begin with the forms. Given listing contracts, sales contracts, disclosure statements, offers, counter-offers, etc., electronic form packages allow for an organized and enhanced workflow in completing documentation within a transaction. Use of electronic form software provides for an automated process of entering in data one-time; with data then replicated across all the electronic forms related to the transaction. For example, the property's street address, seller's contact information, etc., is entered one time then "copied" across forms.

Providers of electronic forms solutions may provide "integration" with various technologies. In terms of consumer and property data, integration with the MLS allows for data to be imported directly from the MLS. Integration with online transaction management tools can provide transfer of forms to such an online repository and transactional interface.

Some providers of electronic form applications now provide online access to their electronic form libraries. This online capability has given the on-the-go agent mobility; providing 24x7 access to print and even email such documentation. Thus, electronic forms and inclusive online access have truly extended the web presence of real estate professionals.

Alternatives to Electronic Forms Software

As it may, electronic forms can be developed with many of today's popular word processing applications like Microsoft Word or document applications like Portable Document Format (PDF). All-in-all, these solutions take time to develop custom forms in-house and are not natively integrated with 3rd party applications like the MLS. To develop form text fields, where data is typed once then replicated across forms like "property address", almost takes a professional programmer to implement properly.

For agents using Word documents or PDF's as their preferred forms application, they must manually enter in data related to the transaction numerous times if replication is not enabled. Such repetitive data entry, "opens the door" for error in data-entry across all forms.

Electronic Forms Software

Providers of electronic forms software are mindful of techniques for real estate professionals to reduce error found with manual data entry. Such providers develop form libraries for MLS/Associations and their members. Or forms providers will create custom libraries for brokers looking to streamline and coordinate forms management within their own firm.

The immediate benefit to implementing electronic forms software and inclusive form libraries is standardization in the "look" of all forms, error-handling in form input, and the ability to minimize redundant data entry. Figure 41, as displayed below, showcases the ZipForm® *Online* Form Viewer. This application organizes forms by libraries, then categories within those libraries.

Case-Study—ZipForm

Through the remainder of this chapter, competencies surrounding electronic forms are complimented by the case-study of ZipForm®.

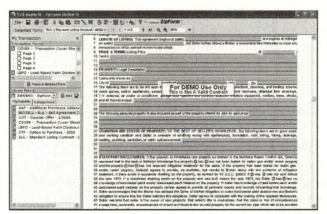

Figure 41: ZipForm® Online Form Viewer—ZipForm.com[90]

It's important to note that each ZipForm® form (as with forms of other providers) is "locked" to prevent inadvertent deletion or modification of the inclusive form text. Rather, as shown in Figure 41, users are only allowed to edit the "shaded" areas (text fields) that denote where a user may type or modify imported text. From a risk management perspective, this cautionary measure as implemented by ZipForm and other forms providers, ensures the "core" document always remains in-tact and cannot be modified on-accident by an agent, assistant or end-user.

Manual Data-Entry

When the majority of form data is manually entered, a forms provider will develop forms where repeatedly mentioned text fields (i.e. sellers name) are keyed by the user one-time. This means that each text block in a form is given a "name". The name acts as a unique identifier, ensuring any reference to the same text field across all pages in forms is auto-populated across all respective forms. Any edit or change to a text box means any occurrence of a similarly named text box is instantaneously updated. What are real estate professionals saying about today's forms software providers?

> "…an important feature of forms management software is the ability to auto-populate the forms with existing data. Only 28% say their forms management software provides this capability. 57%, say they do not have or do not know if they have this capability…" [91]

These numbers are startling that there are some forms providers who do not auto-populate, at minimal, manually keyed data across pages of forms and across forms in a specified category.

---

[90] *ZipForm® Form Viewer*. Retrieved January 08, 2005. RE FormsNet LLC. Reprinted with permission. http://www.ZipForm.com.

[91] *2005 REALTOR® Technology Efficiency Survey*. Center for REALTOR® Technology (CRT). Published May 03, 2005, p36. http://www.Realtor.org/CRT.

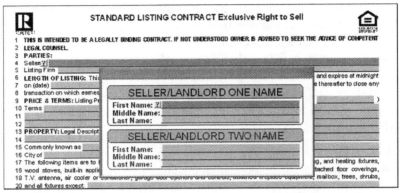

Figure 42: Data Prompt: ZipForm® Form Viewer—ZipForm.com[92]

As shown above in Figure 42 displays the user prompt for the names of a seller within the "Standard Listing Contract". Any page within the transaction that contains the Seller/Landlord names will have that data field auto-populated. Later in this chapter, we'll examine how electronic forms software providers extend the desired auto-populate feature with popular data repositories like the MLS.

Form Libraries

A provider of electronic forms software will categorize broker or MLS/Association forms into categories. Such categories may include: counter offers, disclosures, listing, purchase, lease, commercial, etc. A provider, like ZipForm, may include miscellaneous form libraries that are generalizable of any MLS/Association or brokerage firm. Such forms may include *Lead-based Paint Addendum, BPO (Freddie Mac),* and *Employee Relocation Council (ERC)* for example.

Methods to Access Electronic Forms

Many providers of electronic forms software will provide form viewing and editing with a desktop application, PDA, or even online. ZipForm can do all three!

---

[92] *Data Prompt: ZipForm® Form Viewer.* Retrieved January 08, 2005. RE FormsNet LLC. Reprinted with permission. http://www.ZipForm.com.

Desktop Application

A desktop application of an electronic forms software provides as an immediate repository of transactional documentation in digital (electronic) format. In terms of storing and making numerous copies for record-keeping, such efforts can be reduced or eliminated when using electronic forms software on the desktop. Let's examine ZipForm®*Desktop* as a case-study example.

ZipForm®Desktop Features

- Easy-to-use navigation lets you move quickly through a document or transaction
- Standardized printing—ZipFormDesktop can print to any printer supported by Windows® 2000, ME, XP or NT 4.0.
- Field help buttons allow you to easily identify what transaction information you need for an active field and comprehensive help screens provide instant assistance for important features.
- An autofill & intellicopy feature which allows you to share information between fields using simple point-and-click control
- A clause editor that lets you create, use and store customized clauses
- A complete range of useful tools like spell-check, text strike-out, auto coversheet, auto log-in, amortization scheduling and a mortgage calculator
- Electronic Signatures are available [93]

PDAs and Mobile Devices

Due to the nature of the job, real estate professionals are commonly in the field conducting open houses, visiting with clients and simply on-the-go. The ability to edit and modify electronic forms from a mobile, internet-enabled device is at the forefront of business objectives of the real estate professional. Such ability enables best-practices in time-management to complete work outside the office. The Center for Realtor® Technology has noted this requirement and desire of real estate professionals as stated below.

---

[93] *Go Beyond: ZipFormDesktop.* Electronic Marketing. Kit—ZipForm_GoBeyond.pdf. P4. REFormsNet, LLC. Obtained November, 14 2005. Reprinted with permission.

Realtors® express "...they would like to be able to work on the transaction while they are working at an open house or when they are not in the office..."94

Given moments of idleness when outside the office and away from a PC, real estate professionals want to best utilize their time and get work done. This includes the completion of transactional documentation from anytime and anywhere. Such capabilities can be found when a electronic forms software provider also enables access of forms from a mobile device. Let's examine ZipForm*Mobile* as a case-study example of this desired capability.

<u>ZipForm®*Mobile Features*</u>

- Over 90 fields of relevant transaction data including complete listing information for the seller and the seller's broker and agent, as well as complete sales information for the buyer and the buyer's broker and agent
- A wide range of property information including MLS numbers, list dates and prices, offer dates, addresses, deposits, legal descriptions, tax IDs, inclusions and exclusions
- Real-time edit, save and email of transactional documents stored in a ZipForm*Online*™ account
- And much more!

Online Access

Given the agent resource of wireless-enabled tablet PCs or notebooks and laptops as described in Chapter 14, access to the transaction and electronic forms should not be restricted to the office. With some brokerage firms providing "floor time" to agents on a scheduled basis, access to transactional forms stored on PCs in the office may not be readily accessible.

Rather, transactional forms can be made accessible from any PC having Internet access. Users can access and edit forms, email forms or even print forms when using a forms provider that has an online interface.

---

94 *2005 REALTOR® Technology Efficiency Survey.* Center for REALTOR® Technology (CRT). Published May 03, 2005, p37. http://www.Realtor.org/CRT.

Figure 43: The Many "Faces" of ZipForm®*Mobile*—ZipForm.com[95]

In terms of a brokerage with limited space and supporting many agents and personnel, online access of electronic forms is "smart business"; allowing all professionals to continue progress from any Internet-enabled PC. Let's examine ZipForm*Online* as a case-study example of this desired capability.

ZipForm® *Online Features*

- An intuitive Windows®-based user interface that offers familiar ZipForm features such as clause editor and auto-fill

- Built-in email functionality makes emailing transactions as simple as clicking a button

- Automatic updates ensure that you always have instant access to your up-to-date forms library.

- Template creation tools let you group forms commonly associated with a transaction, further reducing the need to manually input repetitive information

---

[95] *ZipFormMobile*. REFormsNet, LLC. Reprinted with permission. http://www.zip-form.com /products/zipformmobile/productdetails.asp.

• A secure administrative user interface gives broker administrators complete flexibility in managing their offices and users. [96]

Digital Signatures

Enacted by Congress, the *Electronic Signatures in Global and National Commerce Act* (ESIGN) "establishes the validity of electronic records and signatures".[97] From this bill, states have now defined how digital signatures play into business processes within their state.

The ESIGN law, which is technology-neutral, provides general performance based guidelines that eliminates legal barriers to using electronic technology to form and sign contracts, collect and store documents, and send and receive notices and disclosures.

For real estate professionals, the use of digital signatures with electronic forms provides immediate gratification of their use; due to faster completion of an executed transaction.

> *"...28% [of respondents] would like the software to be easier to use. They would like to be able to easily send the information to their clients electronically and would like to incorporate e-Signatures..."*[98]

Use of digital signatures by real estate professionals is a behavioral change in the way many real estate professionals do business. However, it's important to note that consumers may have applied digital signatures to the forms associated with becoming pre-approved for a loan with a bank or lending institution.

When real estate professionals consider the volley of paperwork faxed or mailed back and forth (or re-mailed and re-faxed when misplaced), digital signatures can help streamline and expedite document execution. Considering these documents can be emailed for digital signing, the solution is appealing to Internet-savvy consumers. In working-class communities, consumers in such areas will likely find

---

[96] *Go Beyond the Office.* Electronic Marketing. Kit—ZipForm_GoBeyond.pdf. p5. REFormsNet, LLC. Obtained November, 14 2005. Reprinted with permission.

[97] *Federal Action: Electronic Signatures in Global and National Commerce Act.* © 2005 National Conference of State Legislatures. http://www.ncsl.org/programs /lis/cip/ueta.htm. Obtained November 14, 2005.

[98] *2005 REALTOR® Technology Efficiency Survey.* Center for REALTOR® Technology (CRT). Published May 03, 2005, p37. http://www.Realtor.org/CRT.

the technology appealing, working with electronic forms and digital signatures all mitigated through email. Consider out-of-the-area buyers moving into a new area. Hard-copy forms documentation can get misplaced or can be slow-to-return to the responsible agent when such a consumer is tied-up with many relocation tasks and issues.

Let's face it, consumers check email from work or home more times a day than they do a mailbox or a fax machine. Digital signature capabilities with electronic forms and email combined, appeal to such a consumer.

Given the supportive information above on digital signatures, guiding agents into the era of the "paperless transaction", let's examine ZipForm®*Esign* as a case-study example of this enabling technology.

ZipForm®*Esign*

ZipForm again responds to the desires of the real estate professional and provides its user base access to ZipForm®*Esign*. This service allows consumers to "affix" their digital signature to any document in a ZipForm library. The process includes a Digital Envelop that "contains" a document, where consumers apply their initials or signature. This simulates the manual, hand-written procedure of completing areas on documents that traditionally state "Initial Here", or "Sign Here".

> *"…Collecting an electronic signature is a simple process. If you can send an email, you can send documents for clients to sign electronically…"*[99]

Digital Signatures is a hot topic of discussion amongst risk managers who are actively incorporating digital (electronic) signatures into their risk management procedure & policies. Please consult your broker, risk management official and/or legal counsel on the legalities of using digital signatures as a part of your business process. All-in-all, your consumers will be thanking you for making the transaction convenient and easy to execute; all from any PC with Internet access!

---

[99] *The Dawn of Electronic Signatures*. John Mathers Obtained December 10, 2005. Source: September 2004 issue of National Relocation & Real Estate magazine. http://www.rismedia.com/index.php/article/articleview/7677/1/1.

MLS Integration

Real estate professionals can't count on one hand the number of times they manually type a client's contact information or details of a property during a transaction. When considering transactional documentation, listing outlets of the agent website, Realtor.com, local newspaper, etc., the number of times a real estate professional may key this information can be extensive.

> "...Documentation and forms related to a transaction side can be voluminous and time-consuming. About 39% of the respondents say it takes from 11 to 20 documents to complete a transaction in their market. 23% report that it takes over 21 documents with 7%, reporting that it takes on average over 41 documents to complete a transaction"...[100]

Given the number of documents and forms in a transaction, how many total pages encompass a completed transaction? In one market, as reported by the Center for Realtor® Technology (CRT), storage of transactions has become a problem since each file is anywhere from 200 to 500 pages.[101]

Electronic forms software looks to assist in reducing repetitive data-entry, sometimes eliminating all-together. This is especially true of transactional documentation from listing-to-contract as supported by an electronic forms software provider. This functionality is enabled when electronic forms software can "import" data that is already "keyed" in other data repositories, such as the MLS.

The MLS

When considering forms that solicit a property's mailing address and property characteristics, the MLS is an exemplary repository of such data. However, "cutting and pasting" data from the MLS doesn't cater to the desire to auto-populate fields. As stated in Chapter 9, most MLS providers now adhere to the Real Estate Transaction Standard (RETS). RETS is a "common language spoken by systems that handle real estate information, such as multiple listing services."[102] Let's

---

[100] *2005 REALTOR® Technology Efficiency Survey.* Center for REALTOR® Technology (CRT). Published May 03, 2005, p34. http://www.Realtor.org/CRT.

[101] 2005 REALTOR Technology Efficiency Study. (2004, February 23), WAV Group in conjunction with Center for REALTOR® Technology (CRT). p 17. http://www.Realtor.org/CRT.

[102] *Getting Started with RETS.* Real Estate Transactions Standard (RETS) Working Group. http://www.rets.org/userstart.html. Retrieved 05 Jan 2005,

examine ZipFormMLS-Connect™ as a case-study of interoperability and data exchange of electronic forms software with the MLS.

<u>ZipForm®*MLS-Connect*</u>

The goal of RETS was, "to provide a single interface standard that spurs the development of new and innovative tools for real estate companies and agents to use."[103]. ZipForm®*MLS-Connect* was developed in that innovative spirit; allowing ZipForm® users to import MLS data directly in their electronic forms.

The advantage to direct "data-feed" from the MLS to electronic forms is a streamlined approach in completing documents timely and efficiently considering the reduction of data-entry errors.

Interoperability and data exchange by electronic forms applications with the MLS is imperative to streamlining document completion with minimal error. Integration with CRM tools like Microsoft Outlook® and Top Producer® can provide for effective communication of the forms documentation. Now, native integration with Online Transaction Management (OTM) systems has truly extended the paperless transaction; streamlining workflow for transactional document execution with consumers, responsible agents and peer real estate professionals involved in a given transaction.

Online Transaction Management Perspective

The "teaming" of (1) *Electronic Forms* solutions provided by automated forms management software and (2) *Online Transaction Management* has streamlined the transaction process for real estate agents, peer professionals and consumers. This "marriage" of technologies has facilitated the <u>paperless transaction</u> and meets the requirements for an all-encompassing Transaction Management process as defined by Clareity Consulting. Clareity Consulting defines six core functionalities that encompass quality Transaction Management later described in the next chapter.

---

[103] *Will The Real Estate Industry Standarize Internet Data Exchange?* Greg Herder. Realty Times. http://realtytimes.com/rtapages/20041207_idxstandardization.htm. Published: December 7, 2004. Obtained: December 10, 2005.

ZipForm®*Online* and its simple integration with RELAY™ transaction system, provide seamless export of electronic forms and property data into the OTM-based transaction. As shown below in Figure 44, the entire transaction can be exported to the RELAY™ transaction management system. All property fields within a RELAY transaction are also auto-populated, in addition to seamless import of ZipForm documentation to RELAY.

As stated in Chapter 4, <u>Webography</u> is the art or practice of establishing a *seamless* web presence through the selection of web-based applications, joined together through basic and applied web techniques. Such native integration between both ZipForm*Online* and RELAY is in-concert with the applied techniques of Webography for seamless activity.

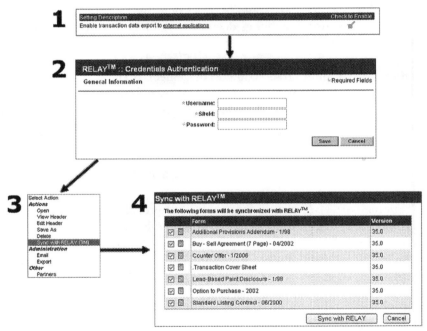

Figure 44: Sync ZipForm®*Online* with the RELAY™ Transaction System[104]

---

[104] *Data Prompt: ZipForm® Form Viewer.* Retrieved January 08, 2005. RE FormsNet LLC. Reprinted with permission. http://www.ZipForm.com.

ZipForm®*Online* and RELAY™ transaction system combine a <u>powerful pairing</u> of technologies that compass all six business objectives to conduct a digital, online, and paperless transaction as stated by Clareity Consulting.

Utility Ordering

As noted in the previous section, OTM systems can be used in-concert with Electronic Forms providers. As with the business and technology relationship with RELAY™ and ZipForm®, ZipForm extends the "customer service" of *service ordering* (as facilitated by OTM systems during a transaction) to *utility ordering* capabilities via its newest service. Utility Ordering systems contacts clients at their request and manages the process of contacting and delivering requests and responses to and from all specified service providers. The solution simplifies the process and is a one-stop utility and service connection service for setup activities for electricity, gas, cable & satellite television, and much more.[105]

Utility ordering (i.e. phone, cable TV, etc) is a value-added extension of the ideal, all-inclusive transaction on behalf of the buyer. It's the nice *send-off* to any buyer who has successfully worked on finishing transactional documentation (closing) while still being provided all-inclusive services from the agent; services to get them best settled in their new home and local community. Such a service gives the buyer "one less thing to worry about" when considering all their relocating activities.

> "…Many agents expressed desire for a method to demonstrate the value they bring to the real estate sales process…"[106]

Utility Ordering is a new service for real estate professionals to put the consumer's interests first and proudly displays the agent's efforts.

---

[105] *ZipFormConcierge: Help your clients connect their utilities in one easy step.* REFormsNet, LLC., Inc. http://www.zipform.com/products/concierge/index.asp. Obtained December 10, 2005.

[106] 2005 REALTOR Technology Efficiency Study. (2004, February 23), WAV Group in conjunction with Center for REALTOR® Technology (CRT). p 28. http://www.Realtor.org/CRT.

Given the supportive information above on Utility Ordering as extending all-inclusive services to the real estate consumer, let's examine ZipForm®*Concierge* as a case-study example of this value-added technology.

ZipForm®*Concierge*

The ZipForm®*Concierge* system contacts clients at their request and manages the process of contacting and delivering requests and responses to and from all of the specified service providers. The solution simplifies the process and is a one-stop utility and service connection service for:

- Electricity
- Gas
- Cable and Satellite Television
- Internet Access
- Local and Long Distance Telephone Service
- Local and National Newspapers
- Financial Services
- Maid Services
- Bottled Water
- U.S. Postal/IRS Change of Address
- And More…[107]

Tips!

When using 3rd party solutions like ZipForm*Concierge* for service ordering, establish an activity in your OTM system that *reminds* you to login to ZipForm*Concierge*, for example, and enter the client and property you wish to submit the "contact request" for service ordering.

---

[107] *ZipFormConcierge: Help your clients connect their utilities in one easy step.* REFormsNet, LLC., Inc. http://www.zipform.com/products/concierge/index.asp. Obtained December 10, 2005.

ZipForm®

ZipForm®, known as the "Official Forms Software of the National Association of Realtors" dates its history back to 1991, as the first electronic real estate forms provider. At present, ZipForm has form libraries available for more than 400 associations and brokers. These form libraries support the nearly 320,000 Realtors® who need to complete a transaction. Forms are commonly navigated and completed using ZipForm®*Desktop*, compatible with CRM tools like Microsoft Outlook and Top Producer, and the MLS via ZipForm®*MLS-Connect*.

ZipForm®*Online* has allowed REALTORS to go "beyond the office". There's no need to carry disks, or mistakenly work with the wrong version of an in-process contract. With ZipForm®*Online*, you can use your laptop, or any computer with an Internet connection, to securely access your up-to-date forms, print them, email them and close the deal.[108]

Activity!

At Webographers.com, candidates of the REAL ESTATE WEBOGRAPHER™ certification work with their inclusive ZipForm*Online* account. Candidates complete electronic forms then seamlessly export those forms to their RELAY™ agent account.

Sales Information
For brokers looking to establish their own custom broker library and special broker account, or for more information, please visit www.ZipForm.com:

- ZipForm®*Mobile*
- ZipForm®*Online*
- ZipForm®*Desktop*
- ZipForm®*Esign*

---

[108] *ZipFormOnline: GoBeyond the Office.* Electronic Marketing. Kit—ZipForm_ GoBeyond.pdf. REFormsNet, LLC. Obtained November, 14 2005. Reprinted with permission.

- ZipForm®*MLS-Connect*
- ZipForm®*Concierge*

Sales inquiries can also be made by calling toll free at 866-MY FORMS (866-693-6767) or their online form at http://www.zipform.com/brokers.

Online Transaction Management

Online Transaction Management technology not only streamlines the transaction's process workflow, which includes documents related to the transaction, but also provides added visibility to *all* the efforts performed by the agent. Such efforts on behalf of buyers and sellers, especially during contract to close, are commonly overlooked and may be underappreciated or undervalued. The transaction, and documentation related to the transaction, has been at the forefront of technology evolution for real estate professionals, their consumers, and peer professionals.

Online transaction management is a behavioral change in how real estate professionals execute a transaction. With OTM, you are changing the way you work, collaborate, how you apply process; making what you do transparent and efficient.

Online Transaction Management: Defined
It has been noted by industry leaders that it's easier to explain the benefits of Online Transaction Management, than to define it. This is simply due to the extensive functionality of Online Transaction Management (OTM) solutions, not to mention the many business use-cases this technology plays a role. Online Transaction Management is an online repository of documents related to the transaction which can be accessed from any Internet-enabled PC, with permission-based access to parties "invited" into the transaction.

	Online Transaction Management	
	Definitions by Leaders in the Real Estate Industry	
Entity	Industry Role	Definition
Center for Realtor® Technology (CRT)	Technology advocate, implementation consultant and information resource.	"...generally includes document and disclosure management, workflow, status tracking, digital signatures, automated service ordering and electronic closing packets..." [109]
Real Estate Business Technologies LLC	Online Transaction Management Service Provider	"...streamlines workflow and collaboration between professionals and clients throughout their real estate transaction..." [110]
Team Double-Click, Inc	Transaction Coordinators, Virtual Assistant Staffing	"The act, manner, or practice of managing, handling, supervising, or controlling a real estate transaction via a secure online environment." [111]
Clareity Consulting, Inc	Consulting for the Real Estate Industry	"An online platform and tool that supports the basic functions listed below as they relate to residential real estate brokerage transactions: 1. Task ('To-Do') Tracking and Management 2. Digital Document Management 3. Participant Set-Up and Security 4. Communication, Notification and Logging 5. Service Ordering 6. Transaction Management (Searching and Reporting)" [112]

---

[109] *2005 MLS Technology Survey.* Center for REALTOR® Technology CRT. Published March 31, 2005. http://www.Realtor.org/CRT, p28

[110] *The RELAY Advantage.* Real Estate Business Technologies, LLC. Obtained December 10, 2005. http://www.rebt.com/Overview.asp.

[111] Welcome to Team Double-Click's Virtual Transaction Office. Team Double-Click, Inc. http://wwwteamdoubleclick.com/virtualtransactionoffice.html. Obtained December 10, 2005.

[112] Transaction Management: A State of the Industry Report. Clareity Consulting, Inc. http://www.callclareity.com/2005-tms.cfm. Obtained December 10, 2005.

How does the rest of the industry best define online transaction management? The table on the previous page looks at various definitions of this competency by many entities involved in the progress and adoption of Online Transaction Management (OTM).

As one can see, Online Transaction Management is technology that can serve many functions. OTM provides the ability to conduct a coordinated and organized transaction made accessible via the Internet. For many real estate professionals and consumers, the online transaction can facilitate the "paperless transaction".

OTM systems can provide organization and secure access of documentation via a corporate and/or personally branded interface. Not only does OTM provide as a central repository for documentation to the transaction, but including branding helps to establish a seamless web presence when made accessible from an agent or agency website for site-visitors. All-in-all there are numerous features commonly found with today's OTM systems to assist in executing an organized transaction, more so than just a document repository.

Caution!

Microsoft Outlook, Top Producer, ACT! are **not** transaction management tools, rather they play the primary role of Contact Relationship Management (CRM). Many OTM systems *may* integrate with these CRM tools.

Traditional transaction management tools include features of: document and disclosure management, workflow, status tracking, digital, signatures, automated service ordering and electronic closing packets. [113]

The Need for OTM

In a 2004 survey of real estate agents conducted by Clareity Consulting, 86% of respondents indicated were either 'interested' or 'strongly interested' in being

---

[113] 2005 MLS Technology Survey. Center for REALTOR® Technology CRT. Published March 31, 2005. http://www.Realtor.org/CRT, p28.

offered a Transaction Management System.[114] Why are the numbers so high? When one looks at the documentation related to a transaction, the number of documents and page-per-documents keeps soaring.

This growing volume of documentation is party due to the risk management efforts to reduce exposure to litigation. Along with the importance of documentation, so has grown the need for better "handling" and organization of transactional documents. The age of the agent keeping a folder of transactional documentation in the backseat of their car is quickly coming to an end.

Forms contain personal identifying information (PII) which requires special handling with respect to privacy. In particular, about 40 states have enacted legislation around PII. There are bills on Capital Hill, at the time of writing this book, regarding such federal legislation.

Agents, transaction coordinators and office assistants are simply inundated with the voluminous amount of transactional paperwork. In one market, as reported by the Center for Realtor® Technology (CRT), storage of transactions has become a problem since each file is anywhere from 200 to 500 pages.[115] Imagine providing hard-copies to all required parties; the amount of ink, toner handling expenses and paper, or even time spent standing next to a printer would be excessive.

Transactions have become extremely complex, disclosures have become a series of documents, and liabilities have risen. Documentation related to the transaction, i.e. disclosures, agreements, inspections, title, escrow, home owners association, and financing can be cumbersome. With various parties involved from the buyer, seller, respective agents, title/escrow, mortgage/lender, inspectors, and other professionals, executing documents has become a "volley" of emails, faxes, and mailings.

Who's Enabling Transaction Management Tools

As of today, we are in the early-adopter phase of the market and life cycle of online transaction management. Also, following the internet bust where there

---

[114] Transaction Management: A State of the Industry Report. Clareity Consulting, Inc. http://www.callclareity.com/2005-tms.cfm. Obtained December 10, 2005.

[115] 2005 REALTOR Technology Efficiency Study. (2004, February 23), WAV Group in conjunction with Center for REALTOR® Technology (CRT). p 17. http://www.Realtor.org/CRT.

about 70 different companies offering OTM, there are about 6-8 survivors today. All-in-all, many real estate agents (today's early adopters) have gone ahead and implemented online transaction management (OTM) for themselves.

In some cases, brokerages have established an agency-wide solution for all their agents and inclusive transactions. In some regions, the MLS/Associations are making Online Transaction Management accessible as a part of member dues. As there is no clear-cut enabler of OTM for real estate agents, agents must "voice" this need to their brokers, who in-turn, need to advocate this need to their MLS/Associations. Or, agents and brokers must take into their own hands, the enablement of online transaction management.

As an interested reader or candidate for the REAL ESTATE WEBOGRA-PHER™ certification you are likely interested in improving your own process; being early adopters with technologies like OTM. Technology implementation looks at adoption by the individual user and can scale up to teams, offices, brokerages, regions served by MLS's and associations.

Benefits of OTM

Online Transaction Management (OTM) is not only an enabler to a more organized execution of a transaction, but provides as a centralized repository of documentation related to each transaction. Again, many of the efforts and activities involved in executing a transaction may be overlooked by clients of real estate agents. There is no better tool that shows the value of an agent, from listing to close, than online transaction management.

Broker/Owner

OTM allows brokers additional "supervision", from listing to closing, of all transactions conducted by agents and transaction coordinators. It provides them a "view" into each transaction, ensuring all activities are on-schedule, timely and organized.

The most important benefit to brokers in implementing OTM is it shows they are taking responsibility to enable their agents with a tool that allows personnel to better maintain activities in a transaction. Brokers are technology "enablers" for their member agents and support staff and must act accordingly.

Agent

Many real estate agents have gone ahead and implemented online transaction management themselves. Agents are natively busy people, spending 50% of their time on activities like scheduling open houses, meeting with clients, responding to phone calls & emails, and more. Winning new clients can be impeded when *not* using OTM for documentation handling, from listing-to-close, with current clients.

OTM can "free up" an agent's time, all-the-while adding another dimension to the services provided to the buyer or seller. OTM allows for the tracking of documentation completed with automatic notifications and mitigating email communication specific to the transaction. OTM commonly requires no additional hardware or software, as found with the RELAY™ transaction management system (the case study application for this chapter) and many other OTM systems. This means that agents are not "tied" to one computer in the office to work through transactional documents. Such documents are made accessible from any Internet-enabled PC, like from the office, home, or mobile laptop.

Peer Professionals

There are some agents who rely on the transaction management software of their title company or 3rd party company.116 Commonly, OTM systems allow for additional persons to be "invited" into the transaction. This entails providing an individual a username and password to access (or upload) documentation relevant to the transaction. Many online transaction management systems allow the responsible agent to categorize these individuals to be on the "Buyer's Side" (sometimes called "Listing Side") versus the "Seller's Side". This ensures particular parties have access just the documents they need to complete their required objectives.

The Consumer

As stated in the Introduction chapter of the book, consumers are extremely web-savvy and comfortable performing daily activities online. Some of these activities include online banking, ordering airline tickets and more. Online transaction management (OTM) accommodates the desires of many consumers who enjoy, and many-of-times prefer, conducting "business" online.

---

116 2005 MLS Technology Survey. Center for REALTOR® Technology CRT. Published March 31, 2005. http://www.Realtor.org/CRT, p29.

In fact, one of your clients may be performing real estate activities online right now. They may be looking for a new home online, or becoming "pre-approved" for a home loan through the website of a bank or lending institution.

Reservations in adopting online transaction management (OTM) is usually with the real estate professional themselves, not the consumer. Always remember what your clients go through in completing all documentation related to the transaction. Many of times they are asked to mail documents or fax when execution should be expedited. With the various parties involved, it may become confusing which party should receive documents from the consumer on a given day. Are they to mail a set of documents to the lender, directly to the title company, or does it go to the agent or parent brokerage? OTM provides consumers one universal interface for completing transactional documents.

Features of OTM

As stated previously, Clareity Consulting defines six core functionalities that encompass a quality Online Transaction Management (OTM) platform. We'll provide a "real-life" platform of RELAY™ that supports the functions of:

1. Task ('To-Do') Tracking and Management
2. Digital Document Management
    (plus integration w/electronic forms solutions)
3. Participant Set-Up and Security
4. Communication, Notification and Logging
5. Service Ordering
6. Transaction Management (Searching and Reporting)[117]

In addition, this chapter will explain how the features of RELAY™, as a case study, best describe the 6 core functions as stated above. The chapter details how OTM systems such as RELAY can also solve many of the concerns as stated by

---

[117] Transaction Management: A State of the Industry Report. Clareity Consulting, Inc. http://www.callclareity.com/2005-tms.cfm. Obtained December 10, 2005.

respondents to surveys of the Center for Realtor® Technology (CRT) during a transaction.

Task Tracking and Management

Tasks and scheduling denotes that the transaction is conducted in an orderly fashion, i.e. before one document or activity can be executed, previous documents and activities must be completed. A solid OTM platform will provide a "checklist" feature, created by the agent or using a pre-defined template. Templates that are repeatedly used for every transaction are commonly defined at the broker level, but sometimes can be defined by the agent user. This ensures "standard procedure" amongst all in the agency (or for the agent's personal transactions), and clearly defines all requirements for a successful and complete transaction.

Project Management Perspective

All agents, transaction coordinators, or virtual assistants who use online transaction management are project managers. Utilization of a transaction checklist that is inclusive of: (a) activities organized by date, (b) "known" dependencies amongst activities, (c) assigned people per activity (responsible agents, consumers, peer professionals), are all the essentials of *project management*.

In terms of project management, your "Critical Path" denotes the time-frame activities are fully completed, from start to finish. Some activities cannot be started until another one is completed. Some activities may be *in-progress* at the same time as others. To expand on this metaphor, Figure 45 on the next page provides a glimpse at the task/activity "checklist" of a transaction found in the RELAY™ transaction system.

Figure 45 on the next page shows how documents are grouped under specific categories like disclosure documents, agreements, title, inspection, etc. The checklist further identifies who is the responsible party for each checklist item, status tracking of the activity, and a desired due date of each activity.

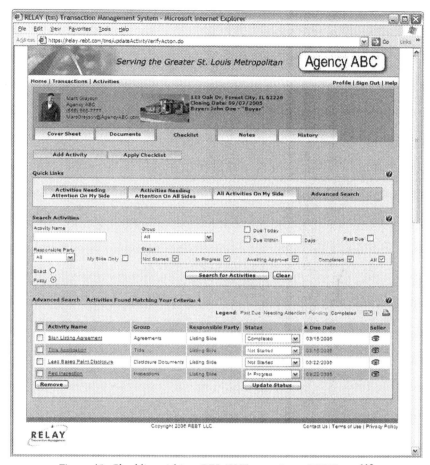

Figure 45: Checklist within a RELAY Transaction—REBT.com[118]

What are real estate professionals saying about the need to better manage the status of documentation and hold responsible parties accountable during a transaction?

> *"…Agents interviewed would like a better method for tracking the status of a transaction. Many study participants said one of their largest frustrations is*

---

[118] *RELAY™ Transaction Management System: Checklist.* Retrieved March 10, 2006. Real Estate Business Technologies, LLC. Reprinted with permission. http://www.REBT.com.

the lack of effective follow-up from their fellow agents, attorneys and lenders. They believe transactions would be closed more efficiently if all parties involved were held accountable even though many believe it would be diffi-cult to get participation especially from lawyers…"[119]

For agents who manage the transaction themselves, no tool is more powerful for organizing documents and managing involved parties than OTM. OTM assists "in-house" transaction coordinators who manage many transactions and need daily insight on the status of every transaction. Later in this chapter, we'll describe how OTM has enabled the use of virtual assistants to assist "power" real estate agents who may be overwhelmed in paperwork. VAs can act as "project man-agers" through the life-cycle of the transaction.

Digital Document Management

Online transaction management natively provides access and upload of docu-mentation in digital (electronic) format. This means your documents in formats of Microsoft Word, Adobe PDF, ZipForm, etc. are securely stored. Exceptional online transaction management (OTM) platforms will remain vendor "neutral".

Faxing

The most dreaded words any agent or real estate professional can say to a con-sumer are, "can you fax it again?" The consumer is left with the questions, "where did the previous fax go?" "Was the document lost?" "Could my private informa-tion been comprised?"

Online transaction management systems, like RELAY, provide an inclusive fax service to help mitigate such issues. This functionality provides the "peace of mind" to the consumer that transactional documentation is handled with care. Inclusive fax services found with an OTM system allow for an inbound fax to be added to the transaction in real-time.

---

[119] *2005 REALTOR Technology Efficiency Study.* (2004, February 23), WAV Group in conjunc-tion with Center for REALTOR® Technology (CRT). p 17. http://www.Realtor.org/CRT.

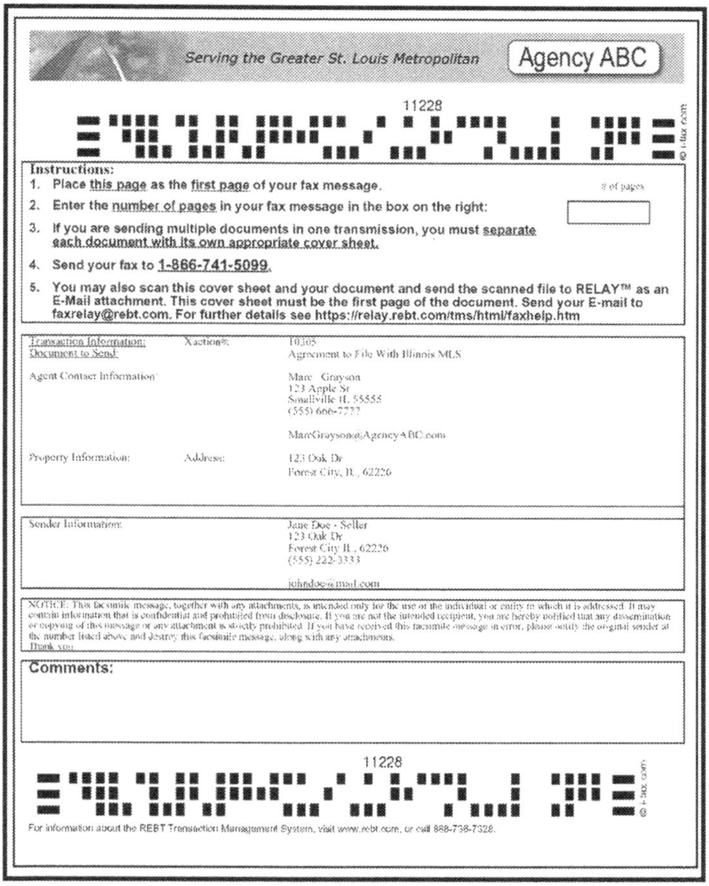

Serving the Greater St. Louis Metropolitan Agency ABC

11228

© i-fax.com

Instructions:

1. Place this page as the first page of your fax message. # of pages
2. Enter the number of pages in your fax message in the box on the right:
3. If you are sending multiple documents in one transmission, you must separate each document with its own appropriate cover sheet.
4. Send your fax to 1-866-741-5099.
5. You may also scan this cover sheet and your document and send the scanned file to RELAY™ as an E-Mail attachment. This cover sheet must be the first page of the document. Send your E-mail to faxrelay@rebt.com. For further details see https://relay.rebt.com/tms/html/faxhelp.htm

Transaction Information:	Xaction#:	10305
Document to Send:		Agreement to File With Illinois MLS
Agent Contact Information:		Marc Grayson 123 Apple St Smallville IL 55555 (555) 666-7777 MarcGrayson@AgencyABC.com
Property Information:	Address:	123 Oak Dr Forest City, IL, 62226

Sender Information:	Jane Doe - Seller 123 Oak Dr Forest City IL, 62226 (555) 222-3333 johndoe@mail.com

NOTICE: This facsimile message, together with any attachments, is intended only for the use of the individual or entity to which it is addressed. It may contain information that is confidential and prohibited from disclosure. If you are not the intended recipient, you are hereby notified that any dissemination or copying of this message or any attachment is strictly prohibited. If you have received this facsimile message in error, please notify the original sender at the number listed above and destroy this facsimile message, along with any attachments. Thank you.

Comments:

11228

© i-fax.com

For information about the REBT Transaction Management System, visit www.rebt.com, or call 888-738-7328.

Figure 46: Fax Cover Sheet—REBT.com[120]

As shown in Figure 46 above, the RELAY™ OTM system, and many other systems, provides the ability for a custom fax cover sheet to be generated for any inbound fax. This cover sheet contains a type of "barcode" which identifies the adjoining document to be a part of a specific transaction. Simply stated, when the OTM fax coversheet is used, the document appears in the Online Transaction

---

120 *RELAY™ Transaction Management System: Fax Coversheet.* Retrieved January 10, 2005. Real Estate Business Technologies, LLC. Reprinted with permission. http://www.REBT.com.

Management (OTM) system within seconds; sending automatic notification of the system's receipt, all-the-while maintaining security and information privacy.

OTM Faxing for Consumers. Returning executed and signed documentation can be a challenge. Whether referring to postal mail or return via fax, OTM systems with inclusive faxing provide numerous benefits and reduce "headaches" in the volley of exchanging documentation to and from responsible parties:

1. <u>Toll-free Number</u>—OTM fax services, as included with RELAY, provide consumers with a toll-free number stated on each fax coversheet. This enables consumers to *readily* return signed and completed documentation from their home or office, knowing that long distance charges are not of any concern.

2. <u>One Fax Number</u>—Consumers can rely on 1 fax number through the life of the transaction when enabled with OTM faxing. This is a huge convenience than being provided the fax numbers of the brokerage, title company, lender, or even the fax machine at an agent's home. With traditional faxing, consumers may be confused what fax number to use for a given document. OTM faxing resolves any confusion with 1 fax number for documents going to any responsible party in the transaction.

3. <u>Fax Goes Through</u>—Brokerages who have limited number of fax machines may have consumers upset when a fax machine is busy. There is nothing more frustrating to a consumer than a busy fax machine when returning signed documentation in a timely manner. OTM faxing services are never "busy" and always ready for fax receipt.

4. <u>Acknowledgement of Receipt</u>. Even when the fax does go through to a specified fax machine, that does not mean it has reached the responsible party. Commonly, consumers "call in" and first state they are faxing, followed by a follow-up call to ensure it has reached a specific individual. When using OTM faxing, a user may "login" to the transaction and see the documents they faxed-in have been "received" in real-time. They don't need to call or email to find the status, but can self-serve that need themselves.

OTM Faxing for Real Estate Professionals. As it may, real estate professionals cannot spend their day standing next to a fax machine. OTM faxing can free up a professional's day, allowing them "peace of mind" that all inbound faxes are indeed reaching their final destination.

1. <u>Knowledge of Sent Faxes</u>—Real estate agents and other inclusive professionals do not have to wonder if a consumer sent a fax. As with the RELAY™ system, agents can set their profile to provide them with automatic notifications. This means any document uploaded or faxed pertaining to them will trigger a notification sent via email. They may also login to the OTM systems and see for themselves if an executed document was faxed.

2. <u>Digital Format</u>—The long-term benefit of Online Transaction Management is that documents are organized in electronic format for enhanced record-keeping. OTM faxing reduces steps in ensuring all documents are in digital format, in-comparison to traditional fax machines.

Tips!

To fully embrace a digital document environment, agents and transaction coordinators can use OTM faxing as an alternative to manually "scanning" documents. Create a fax coversheet from your OTM system and fax all paper documents that are "mailed in" or dropped off in-person directly to your OTM system.

Versioning

As it may, the content of required documentation may change. Or, an agent or transaction coordinator may ask a consumer or peer, real estate professional to make corrections to a form that was done incorrectly. Online transaction management systems allow for versioning of documents held within a given transaction.

This means that a given document with a specific name can have multiple versions stored. An example may include 2 documents that are named the same (i.e. "MLS Listing Agreement.doc"), but are appended with "V_1", "V_2", to denote version 1 and version 2, respectively. This ensures that agents, transaction coordinators, and peer professionals know what copy is the latest version of a given document.

Integration with other Technologies

The overarching goal amongst real estate technology providers is to provide interoperability and data exchange with other real estate applications. Such interoperability may provide the ability to enter in property and consumer information once, then populated to the various technologies used by the real estate professional.

The MLS

Transactions begin with forms. When a property is listed in the MLS, it would be convenient to auto-populate forms with the same property listing information as found in the MLS. A Forms provider who has mastered this feature includes ZipForm® and the ZipFormMLS-Connect™ product. This feature allows ZipForm users to import MLS Data, auto-populating form fields that have not been keyed by the user.

For those looking to integrate the MLS directly with their OTM system like RELAY, should contact Real Estate Business Technologies directly at REBT.com or contact their respective OTM provider or local MLS provider.

Forms Providers

As stated in Chapter 4, Webography asks an individual to look at selecting service providers whose technology may be natively integrated. In today's market, the coupling of a (1) Electronic Forms software and an (2) OTM system should go hand-in-hand. So is true of the RELAY™ transaction system, and ZipForm*Online*.

With one-click integration, all forms from ZipForm*Online* can be uploaded to a RELAY™ transaction, along with all property information displayed in the cover sheet of the transaction. For those who are not ZipForm customers, RELAY can integrate with any application if so desired.

Digital Signatures

Digital signatures, also known as Electronic Signatures, applied to real estate forms and documents provide a faster means for consumers and real estate professionals to execute the transaction. This technology provides a method for consumers to electronically sign or initial, and immediately return the document to the transaction owner. OTM systems, such as RELAY™ are vendor-neutral. As with the case of digital signatures, RELAY™ supports counterpart signatures and digital signing through DocuSign™. Many OTM systems have enabled digital signing technology as a part of their document workflow.

As stated previously, OTM systems may be natively integrated with a forms provider to reduce redundancy in data entry, and provide seamless interaction between the two. RELAY™ and ZIPForm™ continue that relationship with digital (electronic) signatures.

As it may, many electronic forms providers are teaming with online transaction management systems to streamline document execution. For example, ZipForm provides its user base access to ZipFormEsign™, for consumers to affix their digital signature to any document in a ZipForm library. As ZipForm extends this digital signature capability with a Digital Envelop that "contains" a document, along with Stick-eTabs™ where consumers apply their initials or signature.

Given the one-click integration that can be found with an OTM system and Forms provider, as with RELAY and ZipForm, digitally signed ZipForm documents can be seamlessly transferred to RELAY for organization, storage, access and delivery.

Tips!

Real estate professionals and consumers who are not comfortable with digital signatures can utilize OTM fax as a viable solution. Simply create an OTM fax coversheet with every document that requires signature or initials.

They can also scan and email documents directly to an OTM transaction. Both techniques enable the immediate capture and storage of a document, signed and initialed, in digital format.

Create Transaction CD

One of the most powerful features of some online transaction management systems, is the ability to create or "burn" a CD containing all the documentation related to the transaction. As stated previously, the number of documents can range anywhere from 200-500 pages. If there are numerous responsible parties, all whom require copies of documents from a fully executed transaction, this could mean thousands of paper copies.

> "…Providing multiple copies of these folders is also time-consuming
> and expensive. One transaction coordinator, for example, said she
> spends 30—40% of her time making copies…"[121]

The RELAY™ system, for example, allows for the responsible agent or transaction coordinator to create a transactional CD. This provides immense savings in both time and money. Rather than make copies of hard-copy paperwork, "copies" of digital documents are made in seconds.

RELAY™, for example, also allows its users to generate a CD cover that includes branding and identification on the CD for a specified transaction. REBT has taken the time and energy to make the CD experience exactly the same (same user interface, same navigation) as the online interface.

For brokers and risk managers, solid record-keeping is at the forefront of monitored activities within the brokerage. Storing "physical" paperwork in binders is a fading process when enabled with OTM and the creation of transaction CD's.

Simply stated, confined office spaces are beginning to outgrow their storage areas for hard-copy, transactional documentation. In the event of a natural disaster, many agencies store copies of transactional documentation at off-site locations. OTM resolves these issues by storing all transactions for the life of your service. Inclusive search tools of past transactions within an OTM system, and ability to make a copy of transaction documents directly to CD, facilitate all storage and retrieval needs.

Participant Set-Up and Security

Previously in this chapter, it has been stated that OTM provides a *centralized* repository of transactional documents. OTM also provides all responsible parties with a *centralized* interface to all documentation. For many reasons, both practical and legal, OTM systems must also be secure, especially with many invited parties accessing the system over the Internet.

Participants: Buyer's Side vs. Seller's Side

In a given transaction found in an OTM system, both invited users and documentation are commonly defined as being on either the *Buyer's Side* (Listing Side) or the *Seller's Side*. The Seller's side may include the seller, seller's agent, agent delegate (i.e. transaction coordinator, agent assistant), another agent, or any other

---

[121] 2005 REALTOR Technology Efficiency Study. (2004, February 23), WAV Group in conjunction with Center for REALTOR® Technology (CRT). p 26. http://www.Realtor.org/CRT.

individual. The Buyer's Side may include the buyer, buyer's agent, or other individual (i.e. lender, appraiser, etc)

As shown below in Figure 47, users on the Seller's side include the seller, seller's agent, a transaction coordinator who is a virtual assistant from Team Double-Click, Inc. and a new agent from Agency ABC looking to learn the "ins and outs" of a well-executed transaction.

As the owner of the transaction, one can state whether documents added to the transaction are for viewable by the Buyer's side or Seller's side. This ensures that documents are kept private, based on which "side of the fence" the documents are permissible for viewing.

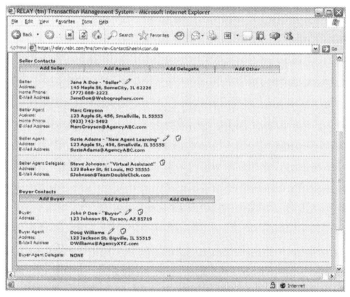

Figure 47: Contacts within a RELAY™ transaction—REBT.com[122]

---

[122] *RELAY™ Transaction Management System: Contacts.* Retrieved January 10, 2005. Real Estate Business Technologies, LLC. Reprinted with permission. http://www.REBT.com.

> **Tips!**
>
> One top producer said he does a transaction debrief at the close of every transaction. "He reviews the entire process with his internal staff, and partner agent. They talk about what worked well and areas where they can improve." [123]
>
> "Training" Accounts for staff during a transaction, or for a "debrief" at the end of a transaction, is a smart training technique. New agents can learn about transactions as a *process*, or an "all-heads-in" review provides open discussion at-close.

Security

One would assume that any online system or website has "security" at the forefront of functionality and procedure. When selecting an Online Transaction Management (OTM) system, security is first and foremost in the minds of REAL ESTATE WEBOGRAPHER™ professionals, followed by functionality.

Why is security important when discussing OTM systems? The Internet is natively open to the world, where technology providers must prohibit unscrupulous site-visitors from comprising user's stored, private information and data. In terms of the transaction, consumer documents and data of the agency must be managed at the highest level of security.

Real Estate Business Technologies (REBT.com)
At the time of writing this book, there is no other highly acclaimed OTM system mindful-of and implementing security best-practices than RELAY™, offered in-part by Real Estate Business Technologies, LLC. Both REBT and its RELAY product received major approvals and high marks for security from two well-respected authorities in security assessment.

1. REALTOR Secure®—best security practices for the overall business organization and administered by the Center for Realtor Technology (CRT). This was awarded to Real Estate Business Technologies, LLC in

---

[123] 2005 REALTOR Technology Efficiency Study. (2004, February 23), WAV Group in conjunction with Center for REALTOR® Technology (CRT). p 25. http://www.Realtor.org/CRT.

2005. As RELAY™ is built by the Realtor-practitioner; its level of security has met rigorous guidelines to protect the practitioner and respective clients.

2. Cybertrust—an Application Security Review, similar to what banks and online trading application go through to ensure that not only are the company's business practices up to snuff, but that the application itself and the operational procedures and practices follow the best in the industry. Both RELAY™ and REBT's security practices passed with high marks in 2005 as based on a Cybertrust review.

When selecting a service provider for online transaction management, pay close attention to the security standards of the application and servicing company. For example purposes, below are specifics on the inclusive security specifications of RELAY™, transaction management system.

- Everything you do on RELAY™ is done through a securely encrypted website, ensuring no one can access your data without your permission.
- Every document uploaded to the RELAY™ system is scanned twice for viruses: once when it is uploaded, and again when it is downloaded.
- Virus definitions on our scanners are updated daily to ensure they remain current.
- Your transaction documents and data are automatically backed up every four hours, and sent to a separate, secure location daily.
- All transaction data is encrypted at a secure, off-site location.
- Temporary usernames and passwords expire if not used within one week.
- You always maintain control over the users allowed into your RELAY™ site.[124]

Communication, Notification and Logging

Some of the most powerful features of OTM systems, include those supporting *Communication*, *Notification*, and *Logging*. Transactions are a dynamic process

---

[124] *RELAY™: Secure Transaction Management.* Retrieved January 10, 2005. Real Estate Business Technologies, LLC. Reprinted with permission. **http://www.REBT.com/security.asp**.

with many responsible parties with many activities and inclusive documents. OTM systems need to be as equally dynamic in supportive functionality.

Communication

It's known that open and organized communication is imperative to any successfully executed transaction. OTM providers like Real Estate Business Technologies and their RELAY™ system has inclusive email functionality throughout the system. The user may desire to email information to a responsible party, or email documents directly from the OTM system. Users do not have to export information or download documents, to later email using their CRM tool like Microsoft Outlook. Such activity can be performed directly from an OTM system like RELAY™.

OTM systems will deliver a message to a recipient you specify, where the email includes you as the sender in the "To:" field, along with your signature block at the conclusion of the email. OTM can streamline email communication, as found with the RELAY system, as an all-inclusive resource to fully execute a successful transaction.

Notification

When using traditional media to execute a transaction like office fax or postal mail, a transaction owner cannot simply sit and wait for these documents to appear. OTM systems that include notification features commonly provide email notification when an "activity" has occurred. Email notification to the transaction owner may occur when a user uploads or faxes (using OTM faxing) a completed document. Or, a seller may receive an email notification when the transaction owner adds a document viewable to the Seller's Side for example.

Again, OTM "frees up" individuals from being tied to one computer, or standing next to the fax machine waiting on a fax. Notification emails help update responsible parties when specific users have uploaded or faxed-in documents to the transaction.

> **Tips!**
>
> OTM systems like RELAY™, found at REBT.com, allow users within a transaction to have a *primary email* and *secondary email* address within their user profile.
>
> Agents, transaction coordinators and other professionals within the transaction can add their mobile email address from their wireless phone provider as their secondary email in their OTM user profile. In terms of mobility, this allows professionals to receive automatic notifications directly to their cell-phone, PDA or any other mobile device.

Logging

Any risk manager will tell agency personnel that all communication and activities during a transaction must be logged. As it may, communication can be electronic via email, over the phone, or through mail.

History

OTM systems like RELAY provide a *History View* of all activities performed with the system. For a given transaction, RELAY logs all activities of when documents where accessed, uploaded, faxed, emailed, etc. In addition, RELAY™ provides a "timestamp" of when these activities occurred and by whom. From a risk management standpoint, nothing provides a better "paper-trail" of activity than the inclusive History, or logs provided by OTM systems such as RELAY™; they represent reality and cannot be altered.

Verbal Communication

Any responsible agent or transaction coordinator understands that logging and tracking documentation is important during the transaction. Verbal communication is as equally important. Agents need to clearly document every communication mechanism (call, email, and meeting) with the homeowner, buyer, respective agents and other real estate professionals. So what are some of the current practices to log verbal communication?

> *"Each agent is handling the documentation of these key issues their own way. Some use a simple "notebook" where they keep all of the details of the transaction. The notebook gets placed in a folder with all of the other documents at the close of the sale. One firm in the study hired a full-time risk management executive dedicated to training agents and working with them to fend off*

arbitration and lawsuits. He is available to handle any "grey" areas and to help agents work through issues." [125]

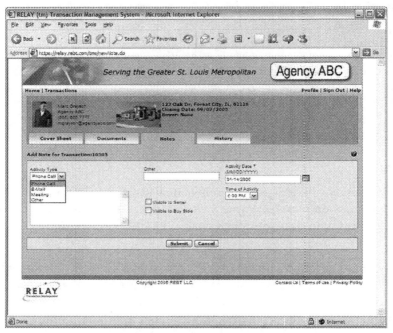

Figure 48: Note-taking within a RELAY™ transaction—REBT.com[126]

Logging verbal communication using a notebook, as mentioned in the citation above, lends itself to many issues. What if the notebook is lost? Are the notes and handwriting legible? Are the notes made in pencil where photo-copies cannot be made?

OTM systems are beginning to embrace this issue head-on, as implemented by Real Estate Business Technologies and their RELAY™ system.

---

[125] 2005 REALTOR Technology Efficiency Study. (2004, February 23), WAV Group in conjunction with Center for REALTOR® Technology (CRT). p 23. http://www.Realtor.org/CRT

[126] *RELAY™ Transaction Management System: Notes.* Retrieved January 10, 2005. Real Estate Business Technologies, LLC. Reprinted with permission. http://www.REBT.com.

As shown in Figure 48 on the previous page, "notes" can be applied to a transaction using the RELAY™ system. Notes can be organized by activity-types of *Phone Call, Email, Meeting,* and *Other*. For each note, *Activity Date* and *Time of Activity* can be applied for a full timestamp of the noted activity. And finally, it can be determined if the Seller (Listing) or Buyer's Side needs access to view the note, or if the note should be kept "private" to the transaction owner.

Service Ordering

Described earlier in this chapter is the ability to manage documents and inclusive real estate professionals using task management capabilities (checklist of organized activities). Such a resource allows for scheduling of activities like "service ordering", to remind the transaction owner, or the consumer, of whom to contact and when for title, pest inspection, hazards disclosure, etc. Note-taking, as described in the previous section can allow for mental reminders to conduct service ordering.

Searching & Reporting

Online transaction management systems will commonly provide a search tool to locate previous transactions. RELAY™ is exemplary of its *extensive* search tools and reporting of transactions. Rather than list *all* the transactions of a given transaction owner or the whole brokerage, RELAY provides category-specific searching to locate the exact transaction you are looking for.

For transaction owners, they may conduct an extensive search via the search categories of:

- Street Address
- City
- Zip Code
- Escrow Number
- Status (Open, Pending, Staged, Inactive, Closed)
- Transaction Owner's First Name
- Transaction Owner's Last Name
- Time Range (between Opening and Closing dates)

The search capabilities as represented by RELAY™ for example allow for *dynamic* searching; as all transactional information is stored in a database for optimal querying. For example, one can perform transaction search on many fields at one time, like:

> Transaction Owner's Last Name = *Pettington*
> Time Range from *Sep 1, 2005* and *February 1, 2006*
> Status = *Open*

The search example allows one to gain results that include transactions that are still open by agent Pettington during the months of September of last year to February of this year. Could you imagine manually searching through hundreds of transactions to find out this information?

Such search tools are great for reporting transactions that may have fallen late or overdue based on your own criteria. In addition, being able to locate historically archived transactions that are similar in characteristics of a current transaction being executed, may provide best practices or good research.

Transaction Coordinators

As stated earlier, transactions have become extremely complex, Documentation is voluminous, i.e. disclosures, agreements, inspections, title, escrow, home owners association, and financing. Keeping order amongst various parties from the buyer, seller, respective agents, title/escrow, mortgage/lender, inspectors, and other professionals, is also a challenge.

OTM is an excellent enabler for transaction coordinators to best execute a transaction. In-house transaction coordinators will thank you whole-heartedly for implementing the solution to assist in their daily efforts in managing transactions agency-wide.

> *"Throughout the early days of a listing, the listing agent is working to complete disclosure documents that need to be included in a closing packet. In one market in the study disclosures are significant so this process can take some time to complete. One firm in the study has employed transaction coordinators to help agents handle the paperwork required to close a sale. Many top producers hire assistants who handle most of the processing required. This*

allows them to focus more time on generating leads and listings and provid-ing superior personal service to their clients. [127]

As described earlier, transaction management is rooted on the principles and fun-damentals of project management. In many agencies, a "project manager" is required due to the extensive activity and oversight required for each transaction. Such managers may be an in-house transaction coordinator or a personal, virtual assistant for power-agents.

Virtual Assistants

Real Estate Business Technologies, LLC has left no stone un-turned in developing an Online Transaction Management (OTM) system on the market that is mind-ful of the practitioner. Sometimes, technology is not always about the latest and greatest features, it's about the people who use the technology. Real Estate Business Technologies not only provides training for those desirous of being a RELAY™ Certified Trainer, but makes these individuals accessible to its cus-tomer based at http://www.rebt.com/TrainersList.htm#National.

The pairing of virtual assistants and online transaction management has a long history. OTM truly enables an assistant to facilitate the execution of a transaction without physical contact with the responsible agent.

Team Double-Click

For those top-producing agents or brokers looking to obtain transaction coordi-nators, Real Estate Business Technologies at REBT.com has that covered for you. REBT looks to ensure that transaction management enables top-producers to focus on getting more sales.

Team Double-Click, Inc. a strategically aligned partner of REBT, provides pro-fessional real estate, virtual assistants for small and home-based businesses. Their transaction coordinators and agent assistants help streamline your transaction with the use of RELAY™ OTM system. For more information, please see http://www.teamdoubleclick.com/virtualtransactionoffice.html to learn about obtaining VA support, powered by the RELAY system.

---

[127] 2005 REALTOR Technology Efficiency Study. (2004, February 23), WAV Group in conjunction with Center for REALTOR® Technology (CRT). p 22-23. http://www.Realtor.org/CRT.

Personal & Corporate Branding

Online Transaction Management systems are an extension of the agent's, broker's, or MLS Association's web presence. As many brokers are currently enabling OTM for their inclusive agents and in-house personnel, OTM systems commonly include "personalization" features to make the system "your own".

> *"Agents did express an interest, however, in a way to keep a client constantly updated during the transaction via a personalized website which would be linked to a transaction status report. In addition, they would like to maintain the website and use it as a way to maintain a relationship with their customers."*[128]

Figure 49: RELAY™ : Corporate and Personal Branding—REBT.com[129]

RELAY™ transaction management is a good example allowing *corporate* and *personal* branding to establish a seamless web presence. As shown in Figure 49 above, site managers for the brokerage can apply a "top banner" that is displayed on all transactions, seen by all users. Personal branding allows the agent to add a photo, showcasing they are the "supervisor" of this transaction, and take ownership of its successful execution. This personalization supplements the "human touch" found in a paper transaction during the digital transaction.

---

[128] 2005 REALTOR Technology Efficiency Study. (2004, February 23), WAV Group in conjunction with Center for REALTOR® Technology (CRT). p 29. http://www.Realtor.org/CRT

[129] *RELAY™ Transaction Management System: Notes.* Retrieved January 10, 2005. Real Estate Business Technologies, LLC. Reprinted with permission. http://www.REBT.com.

OTM—Link Everywhere

It's recalled from Chapter 4 that webography is the art or practice of establishing a *seamless* web presence through the selection of various web-based applications, joined together through basic and applied web techniques. Showcasing an OTM system in an agent or broker's web presence is perfected through webography.

OTM incorporated into an agent's web presence feeds the desire of consumers to "self-serve" their execution of documentation. Making your OTM service known to consumers from your web presence, displays you're a one-stop, all-inclusive service provider for both sellers and buyers.

An agent should make a point to reference the URL of their corporate or personal OTM web page in every meaningful outlet of their web presence. Such outlets in an agent's web presence where the OTM web page should be referenced for consumers and peer real estate professionals includes: the agent website and single property websites.

Agent Website

Commonly, the URL of a RELAY web page will look like: https://relay.rebt.com /tms/?siteId=10168. The URL is specific to the agent, broker, or MLS subscriber, where consumers and site-visitors are directed to a corporate or personally branded login page.

As showcased in the RELAY™ 2005 Winter Newsletter, tips are provided to best hyper*link* your OTM web page for clients at http://www.rebt.com/RELAY _Newsletter_Dec_05.asp. As with Figure 50 on the next page, you see that Greg Dugan, a top-producing agent, added a custom navigational button (on the left-hand side) to identify where to access the OTM system. In addition, he created a custom-page that describes the service he is providing to sellers, buyers, and peer real estate professionals. This "splash page" is an effective technique to describe your all-inclusive services, which includes OTM, to site-visitors, current and potential clients.

The above best practice holds true for any online transaction management system; creating an intermediary web-page between the agent website and the OTM login page. It allows the agent or agency to boast about the all-inclusive services provided in addition to explaining OTM's overall purposes to site-visitors.

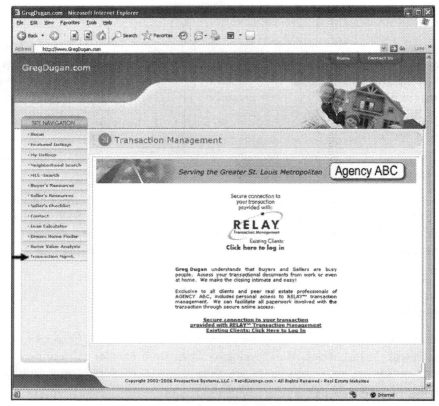

Figure 50: Agent Website with RELAY™ Access

Greg Dugan took efforts to incorporate the same agency banner that appears on the RELAY™ transaction management system in the splash page. As stated in Phase 4 of Webography, *Establish a Seamless Web Presence*, the process notes using applied techniques to "transition" site users across applications.

In addition, Greg included two hyperlinks on the OTM splash page that redirects users to his agency's OTM webpage. One hyperlink includes the "RELAY" logo at the top of the splash page, followed by the hyperlinked text at the bottom of the page. Both techniques are described in the RELAY 2005 newsletter exclusive to RELAY subscribers! Once made available on an agent's or broker's website, no additional maintenance is required on behalf of the agent.

Single-property Websites

With any AgencyLogic.com PowerSite (i.e. www.123AnySt.com), custom "menu" buttons can be added to the navigational bar.

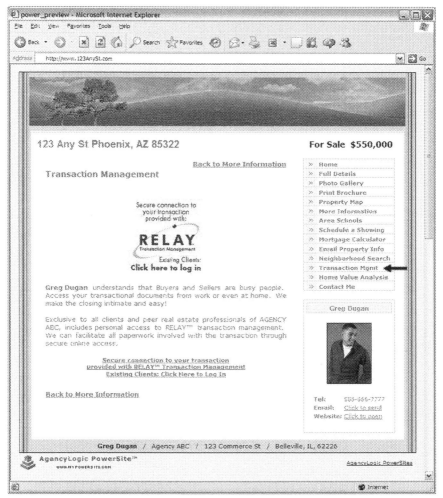

Figure 51: Single-property Website with RELAY™ Access

The buttons can take a user to a webpage you created in the AgencyLogic control panel, or act as a hyperlink to an external web page. Just like the agent website, an OTM "splash page" can be established as a web page you create. When presented in each AgencyLogic PowerSite, it enables an all-inclusive, single-property web-site. This process is displayed on the previous page in Figure 51.

In conclusion, OTM is a powerful tool to organize the workflow of transactional documentation. Online Transaction Management is an extension of an agent or brokers web presence. OTM provides consumers the visibility into the closing process. Such efforts may have been previously overlooked, but now admired through Online Transaction Management™.

REBT.com

RELAY™, a service of Real Estate Business Technologies LLC located at www.REBT.com, is the Online Transaction Management system designed by the REALTOR® practitioner. Hailed by industry leaders as a mission critical enhancement to the services provided by REALTORS® to homebuyers and sellers, RELAY™ allows the tracking and management of all information related to a real estate transaction from listing through closing; allows users to enable assistants, other agents or transaction facilitators to participate in the process; and features one-click integration with ZipForm® and WINForms Online®, the electronic forms software used by nearly 340,000 REALTORS® nationwide.[130].

Sales Information

For information on establishing agent or broker accounts, please view Real Estate Business Technologies website at www.REBT.com or for all sales inquiries email info@REBT.com or call directly at 1-866-736-REBT.

---

[130] *The RELAY™ Advantage.* Real Estate Business Technologies, LLC. Retrieved December, 10 2005 http://www.REBT.com/RELAY.asp. Reprinted with permission.

PART SIX:
On-the-Go and in the Office

Chapter 14: Mobile Technology

Chapter 15: Additional Hardware & Software

Mobile Technology

Notebook PC's (laptops), PDA's (Personal Digital Assistant), and SmartPhones have given mobility to the real estate professional. One can work on current tasks from any location, provide live demonstrations & examples to potential clients, or showcase previous work-performed to current clients.

Mobile technologies go beyond standard operations like email, text messaging, and calendaring performed on a laptop or PDA. Imagine having virtually everything from your office desktop computer and having it available on a mobile device. Consider activities on your PDA such as: search and access MLS data and virtual tours, review documents and much more. Such activities are paramount to conducting business while on the go, impressing your clients at in-person meetings with mobile technology.

Let's first examine some of the hand-held technology that exists in today's market to see how real estate activity can be executed while on-the-go.

Mobile Devices

Notebook PC's (laptops) and PDA's (Personal Digital Assistant) add mobility to a REAL ESTATE WEBOGRAPHER™ professional. One can work on current projects from any location, provide live demonstrations & examples to potential clients, or showcase work-performed to current clients. Such mobile technologies allow real estate professionals and independent contractors the ability to work from any location. One must examine their existing mobile technologies and examine the possibilities to upgrade or modify a current toolset.

SmartPhones

A SmartPhone is cell phones that can provide an enhanced color screen with a robust operating system. Commonly the screen is about 2.25 x 2.5 inches, 62 colors, and TFT resolution. This piece of hardware is useful in conducting basic operations like checking email and writing short email responses, or following up to the message w/a phone call.

A drawback to using smart phones for business transactions, like email, is the keypad is commonly numeric-only. Just like the key pad on any touch-tone phone, you must press the numeric key one-to-many times to reach the desired letter. This can become tedious and time consuming.

However, manufacturers like Samsung have provided a touchpad that uses a stylus pen to "write" to resolve the issue of a tedius typing by numeric keypad. These phones, like the Samsung SPH-I500 can be very affordable, especially if you enter in a long-term contract with a cell phone service provider that offers such a smart phone. Although these phones may appear small and limited in functionality in comparison to a notebook PC, the tool provides for quick communication with clients or customers who demand timely information and acknowledgement through immediate email messages and phone calls.

Personal Digital Assistant (PDA)

PDAs provide the ability to perform many functions of a personal computer (PC), with the convenience of carrying something light and unobtrusive when you're on the go! In the real estate market, taking a phone call, responding to email, checking your calendar can be critical. PDA's with wireless technology allow you to perform all those functions in real time.

When searching for a PDA, you must ask yourself: do I want to replace my existing cell phone for an all-in-one PDA (phone, Internet, and PDA)? To gain Internet access with non-phone PDA's, additional hardware may have to be purchased. Wi-Fi SD Card is another add-on to many PDA's which allow for users to access a wireless network or any wireless hotspot. Such activity is similar to having a notebook PC (laptop) and accessing a wireless network.

For all-in-one PDA's, Internet access can be established through your cell phone service provider. Many of the major cell phone service providers have partnered with manufacturers of PDA's to provide such all-in-one capability. It's best to start with your existing service provider of cell phone service or look to a new one to make such a purchase of an all-in-one PDA.

Included with a PDA is a cable (commonly USB) and software to "hot sync" functions with a personal computer (PC) or notebook PC. This allows one perform functions on the PDA when offline, but when synced with the PC, pre-composed email messages for example are sent. Some PDA's have a touchpad which requires a stylus pen to simulate "writing". Others like the palmOne Treo 650, has a Directional Pad for navigation on the screen and a keyboard to simulate typing.

Notebook PC

A Notebook PC (laptop) and a tablet PC should have comparable specifications as that of the personal computer (PC) as described earlier in this chapter. The main evaluating factors with Notebook PC's may include weight, screen size, operating system, and included software (i.e Microsoft Office) battery life, integrated 802.11B (wireless). Below are some addition specs to consider when purchasing a new notebook PC.

Featherweight PCs (thin and light) are great when on-the-go, but may lack internal bays for optical drives (DVD-CDRW) floppy drives, external hard drives, etc. These drives are commonly external with featherweight PC's and include additional costs. Larger Notebooks can provide for inclusive drives like floppy and optical drives and are usually bundled into the overall price.[131]

In terms of the web-based technologies stated in previous chapters, many have features that facilitate mobile device activity. Lets take a look at previous technologies mentioned in this book and see how they facilitate the on-the go agent.

Email Notifications from Web-based Technologies

Resources like a BlackBerry® allow users to access email sent to fany specified email address from a mobile device like a wireless PDA.[132] For agent's who utilize

---

[131] Heavy Hitting Featherweights. PC World Magazine online. Obtained June 5, 2005. http://www.pcworld.com/reviews/article/0,aid,86950,pg,2,00.asp
[132] BlackBerry® is a registered trademark of Research In Motion Corporation

smart phones or wireless PDA through their cell-phone provider, what can agent's do to receive instant notification to their hand-held of notification emails from their web presence?

Many of the technologies discussed in this book allow agent users to establish a secondary email address in their user profile. When provided an additional email address from your cell phone provider, like user12345@somecellphoneprovider.tld this can be a powerful resource to receive 'lead" emails from your website, or other web form generated emails.

An example may include Greg Dugan establishing his primary email in his RapidListings.com, agent account to be Greg@GregDugan.com; whereas his secondary email is user12345@somecellphoneprovider.tld. Since email notifications are sent to both email addresses, if Greg is out of the office, he can be notified immediately of a potential lead generated from his website for example. As stated in Chapter 5, site-visitors expect a timely response from agents when they use submission forms on an agent's website or direct email. Such a technique will assist in returning information timely to the requestor.

Mobile Applications Unleashed

Never before seen in the market place for agents now comes Mobile Technology that goes beyond simply checking email or your daily calendar. Imagine having virtually everything from your office desktop computer and having it available on a mobile device. Consider activities on your PDA such as: search and access MLS data and virtual tours, review documents and much more.

Case-Study
iseemedia and its North American partner RealBiz360, have taken great strides to provide extensive mobile applications for the real estate professional. Iseerealty, developed by iseemedia, provides access to MLS data, virtual tours, documents and more on your wireless-enabled PDA or SmartPhone. As the distribution partner for North America, RealBiz360.com will make this technology available to its customers in late 2006.

iseemedia

iseemedia, Inc's iseerealty™ is a mobile client server application using the iseemobility platform to deliver a must have application for all real estate agents.

It provides four main areas of functionality: MLS data access; real estate applications; access to online Virtual Tours; and access to online documents. The application is designed to provide the agent with a strong complement of sales, marketing, communications and customer management tools to make the mobile device as effective as their desktop computer at the office or home. iseemedia has developed a Real Estate virtual tour creation and publishing portal with partner RealBiz360.com. iseemedia's extensive research and development has reached the real estate market place via RealBiz360.com.

iseerealty™

iseerealty is a mobile client server application using the iseemobility platform to deliver a must have application for all real estate agents. It provides four main areas of functionality: MLS data access; real estate applications; access to online Virtual Tours; and access to online documents. The application is designed to provide the agent with a strong complement of sales, marketing, communications and customer management tools to make the mobile device as effective as their desktop computer at the office or home.[133]

Search

- quickly search for virtual tour and MLS listings
- use advanced search features to limit searches to specific property attributes
- save searches and recall them in seconds
- iseemedia smart client technology keeps you working over intermittent networks

Inform

- view title searches, escrow agreements, surveys
- view local cultural data (for example, proximity to schools and parks)
- locate the property in the neighbourhood

---

[133] *iseerealty—Access MLS data, virtual tours, documents and more on your phone enable PDA or SmartPhone.* Iseemedia, Inc. http://www.iseemedia.com/main/products/isee-realtor. Retrieved December, 10 2005. Reprinted with permission.

- send documents to your clients for secure viewing
- send links to your clients for virtual tours
- manage client relationships from your mobile device
- take pictures of homes with your camera phone and send to your clients now

Choose

- help your clients decide while on the road, no need to spend time traveling back to the office
- one click and you can place a call, send an email or SMS to the listing agent
- take advantage of the iseemobility uptime to arrange tours and view homes when the sellers are not accepting visitors
- iseemedia has developed a Real Estate virtual tour creation and publishing portal with partner RealBiz360.

This www.realbiz360.com portal is accessible to all agents, brokers and soon photographers in North America. By providing an easy to use web infrastructure for agents to create virtual tours and publish them online, iseemedia is also facilitating the demand for virtual tour content creation. Virtual tours are delivered to the web and to mobile devices by the iseemedia ImageServer platform for mobile devices. [134]

MLS Data Access

QuickSearch—Search MLS by ID or address—provides online access to MLS Data. The agent inputs an MLS ID number or a property address, and receives the MLS information on the mobile device.

Extended Search and CRM—Create Custom Searches based on various criteria and store by client.

---

[134] *iseerealty—Access MLS data, virtual tours, documents and more on your phone enable PDA or SmartPhone.* Iseemedia, Inc. http://www.iseemedia.com/main/products/isee-realtor. Retrieved December, 10 2005. Reprinted with permission.

Notify—Send notification if property that matches criteria becomes available.

Map Access—Show listing on a map.

| MLS Data Access | Agent Tools | Virtual Tour Viewing |

Figure 52: The Many "Faces" of iseerealty™—IseeMedia.com[135]

Agent Tools

Document Viewing—Access or download Microsoft Office documents such as Word, Excel, etc. Access or download floor plans, surveys or any other legal document for review.

Calculators
- Mortgage—Standard mortgage calculator with updatable rates
- Loan—Calculator that provides loan amount and monthly payment based on household income and expenses
- Home Equity Calculator
- Loan Balance Calculator

---

[135] *iseerealty—Access MLS data, virtual tours, documents and more on your phone enable PDA or SmartPhone.* Iseemedia, Inc. http://www.iseemedia.com/main/products/iseerealtor. Retrieved December, 10 2005. Reprinted with permission.

Statistics—Total Properties Matching Criteria—This function is designed to give the agent and client an indication of general home price trends based on location and other variables.

Virtual Tours Access

Search for a Tour

- Quick Search—enter an MLS ID number or address and display the corresponding virtual tour.

- Advanced Search—Enter property criteria and search the virtual tour database for all matching properties.

View Tour—Tour can be stored locally and viewed on the mobile device when not connected.

Remote—a list of selected tours has been stored locally and the tours are available for interactive viewing online

Save Tours—Save individual tour or search results and associate with client.

Share Tours—Send tours to clients using email or MMS.[136]

Sales Information

For sales information on establishing RealBiz360 Mobile (powered by iseerealty), visit www.RealBiz360.com, call toll free at 1.888.REALBIZ (732-5249) ext. 81 to speak with a sales representative, or email sales@realbiz360.com.

---

[136] *iseerealty—Access MLS data, virtual tours, documents and more on your phone enable PDA or SmartPhone.* Iseemedia, Inc. http://www.iseemedia.com/main/products/isee-realtor. Retrieved December, 10 2005. Reprinted with permission.

Chapter

15

Additional Hardware & Software

Many of the tools to enable a REAL ESTATE WEBOGRAPHER™ to complete online business objectives may already be in inventory. Whether you look to self-serve all your online business endeavors yourself, or enable an "in-house" assistant, included in this chapter are additional hardware and software resources that you should have readily available. For some of the suggested items, you may find them as required or optional based on your business objectives you have defined in the Webography process; the guide to online success as stated in Chapter 4.

Hardware

This chapter focuses on web site capabilities that are commonly found in web sites for the real estate market. This section hopes to bring awareness to capabilities that even a seasoned web designer may not be familiar with. Also discussed in this chapter include web development tools that are required for the success of any REAL ESTATE WEBOGRAPHER™ professional.

Personal Computer (PC)

Although an implied and understood tool of a REAL ESTATE WEBOGRA-PHER™ professional, it's important to take a look at the personal computers (PC). Many of today's basic personal computers currently in retail, have enough "horse-power" to handle the requirements of a REAL ESTATE WEBOGRAPHER™ professional. Considering the purchase of a new computer or looking to upgrade your 5 year old computer? Below are *minimal* specifications for a personal computer:

- Processor: 1.6 GHz
- System Bus Speed: 400 MHz
- RAM: 512 MB DDR SDRAM (Max. 1 GB)
- Cache: 512 KB
- Hard Drive: 40 GB
- DVD/CD-RW: 48x (read), 48x (write), 24x (rewrite)
- Network Connection: 10-/100-Mbps Ethernet
- Graphics Card: Up to 32 MB shared memory
- Operating System: Windows XP Home Edition, Mac OS
- Floppy Drive: 1.44 MB, 3.5-inch floppy disk drive
- Ports: 2 USB, 1 Parallel, 1 Serial, 2 Free PCI Slots

Recommended Accessories/Services:

- Thumb drive-128 MB
  (carry around pertinent files to share w/clients)
- High Speed Internet—T1/DSL/Cable
  (See Chapter 5 for more details)
- Word Processing Suite—i.e Microsoft Office.

Digital Camera

Photos taken with a digital camera lend themselves to integration and compatibility with the web, more so than a standard film camera. Although such photos can be less in resolution than film cameras, the quality produced from digital cameras is sufficient for web purposes. The quality is so sufficient that many-of-times its quality must be manually reduced. When photos are taken with a digital camera, the photos are in digital format, i.e. JPEG, which are conducive for fol-low-on editing for the web.

For web design in the real estate, minimal requirements for a digital camera include:

- Resolution: 3.0 Mega Pixels
- Optical zoom: 3 X
- Digital zoom: 2.5 X
- Max focal length: 18 mm

- Min focal length: 6 mm
- Focus modes: automatic
- Exposure settings: program, automatic
- Image format: JPEG
- Interface: USB
- Flash Media Card Compatible

Recommended Accessories:

- Memory: Flash Media Card (256 MB)
- Tripod with panoramic head
- Camera Software (*drivers, photo-editing, etc*)

Number of Images per Flash Media Card*									
Camera Type	File Size	16 MB	32 MB	64 MB	128 MB	256 MB	512 MB	1GB	2GB
2 Megapixel	900KB	17	35	71	142	284	568	1137	2275
3 Megapixel	1.2MB	13	26	53	106	213	426	853	1706
4 Megapixel	2.0MB	8	16	32	64	128	256	512	1024
5 Megapixel	2.5MB	6	12	25	51	102	204	409	819
6 Megapixel	3.2MB	5	10	20	40	80	160	320	640

Camera Lens

If you're considering taking wide-angle photos, consider adapting a wide-angle lens. Commonly, people have 35 mm cameras, your typical point and shoot. Many vendors provide lens adaptors that allow you to take wide angle, 28 mm photos. There are adaptors for some cameras, or you can invest money in wide-angle digital camera to capture more of a "space" with one shot.

Lens converters expand the cameras filed of zoom. The lens converter is place on the front of the existing camera lens. Sometimes you may need a ring adapter to

---

* Average numbers can vary due to make and model of the digital camera and resolution of the outputted JPEG image.

ensure proper fit of the additional lens piece. Since digital cameras come in all shapes and sizes, in addition to lens converters and ring adapters, its best to contact the manufacturer of the model you currently own.

Tripod/Panorama Head

When taking still photos or especially when taking photos to generate a virtual tour, tripods can be a helpful accessory. In terms of virtual tours as discussed in Chapter 8, a tripod helps steady a camera, taking photos across even lines to ensure proper overlap of photos for stitching and finally outputting a tour. As virtual tours has become a necessity with every property listing, below are some additional accessories that work in-concert with a tripod

Rotating Panorama Head
A rotating panorama head allows the swivel of a camera across an even line. Panoram Head should include: sisks to accomodate various tripods, bubble level, bracket for Fisheye, even mounting surface.

Bundled Kits
There are many vendors of bundled kits that include tripod, panorama head, carry case and even a digital camera. To eliminate time in investing compatible hardware, bundled kits provide a set that will all work seamlessly together, eliminating the guesswork involved with purchasing.

Tips!

VirtualTourWebStore.com (a "sister" service of RealBiz360.com) provides digital cameras, tripods, tripod heads, carrying cases and bags for professional photo-takers. In addition, they provide "bundled" packages for one to purchase an "integrated" kit; perfect for yourself or an assistant.

Scanner

It's handy for a REAL ESTATE WEBOGRAPHER™ professional to have a scanner as a piece of readily available hardware. Too often, images required for a web presence may be currently in print format, not digital format. For a REAL ESTATE WEBOGRAPHER™ professional to provide a timely solution, it's important to have a scanner readily available. Minimal specifications a REAL ESTATE WEBOGRAPHER™ professional should look into for purchase of a scanner include:

- USB 1.1 or 2.0 interface
- 600x1200 dpi optical resolution
- Compatible with the Windows or Macintosh operating systems

Software

On your personal PC, you may have readily available software like Microsoft Office and other applications pertinent to your business as a real estate professional. Below are some additional software applications and services to consider when conducting business within your own personal web presence.

Portable Document Format (PDF) Document

It's important for a REAL ESTATE WEBOGRAPHER™ professional to fully understand how to display documents in formats that are considered "universal" to users. In the real estate market, this could be a brochure of the home-for-sale (CMA) and property's features and amenities. Portable Document Format, commonly known as PDF is the suggested format for such documentation.

PDF converter applications
- PrimoPDF (free)
  http://www.primopdf.com
- Adobe Acrobat 7.0 (one-time cost)
  http://www.adobe.com/products/acrobat/main.html

As-needed-basis
- Adobe Online (pay-per-conversion)
  http://createpdf.adobe.com

Digital Imaging Editor

Whether discussing the editing of photos taken with a digital camera, images outputted by a scanner, or designing original images & artwork, a digital imaging editor is required. It's important to choose a digital imaging editor that works well with your web site development editor discussed later in this chapter. Also, a digital imaging editor provides a platform for custom imaging for brochures advertising, one's REAL ESTATE WEBOGRAPHER™ professional services or other desktop publishing needs.

A work-able digital imaging editor lends itself to assisting those who may not be artistically inclined. Even if you don't know much about graphic design, a digital imaging editor will assist in editing digital images, and tweaking basic artwork unique for that site. Also, simple factors like web-safe colors, are usually presented to a designer in a digital imaging editor, and provide additional tools to ensure images are web-friendly. Some popular digital imaging editors include: Adobe Photoshop® 7.0[137], Macromedia *Fireworks®*[138] *MX*, and Jasc's *Paint Shop Pro®*[139].

Stock Photography

As a REAL ESTATE WEBOGRAPHER™ professional, it is important to always create a perception of professionalism on a client's web site. Consider many corporate web sites out there in existence. On such a web site, you may find a picture of a secretary with a headset on the "contact" page. That person in the picture may not even work for that company, but provides for a professional appearance! For the market of real estate, a professional feel can be established through stock photography geared towards residential and commercial real estate.

Stock Photography in a Box
As a REAL ESTATE WEBOGRAPHER™ professional, it may be beneficial to have a suite of stock photography on-hand. Online vendors may provide purchase of a CD which contains a collection of photos based on a given theme, like the IT industry. Or a REAL ESTATE WEBOGRAPHER™ professional may want a variety or collection of stock photos they can buy at a store. Popular sources of stock photography and images you can purchase at your closest software store include *Photo-Objects®* and *The Big Box of Art®* series from Hemera Technologies, Inc.[140]

Online Vendors of Stock Photography
There are many online providers of stock photography, where some charge for: (1) membership for full access to all pictures for a given time-frame, (2) provide

---

[137] Adobe Photoshop® is a registered trademark of Adobe Systems, Inc.

[138] Fireworks® is a registered trademark of Macromedia, Inc.

[139] Paint Shop Pro® is a registered trademark of Jasc, Inc.

[140] Photo-Objects® and The Big Box of Art® are registered trademarks of Hemera Technologies, Inc.

for orders of CD's containing a selection of photos in a given theme, or (3) charge per photo for download. There are three types of stock photography that can be purchased by an end-user: Flat-Rate, Rights-Managed, and Royalty-Free.

<u>Flat-Rate</u>

Flat-Rate stock photography is lower in price, but the end-user is allowed to use the image one-time. For example, once a flat-rate stock photo is purchased, the end-user may use it for a web site, but can't use the same image for a brochure.

<u>Rights-Managed</u>

Rights-Managed allows the vendor to specify a price for the photo based on the scope of how it's used. For example if the image is invoiced for a print-ad like a brochure, the image cannot be used for the web because the price is based on the agreed usage.

<u>Royalty-Free</u>

Royalty-Free stock photography include photos where after purchase, you do not have to pay royalties. Royalty-free should apply whether using the photo in print work (i.e. brochures) or for web sites. You can use such photos in any manner you would like and as many times as you'd like. Royalty-free images are the suggested type of stock photography for REAL ESTATE WEBOGRAPHER™ professionals. These photos are more expensive than those that are not royalty-free. A simple online search should provide numerous on-line vendors of stock photography.

Quality of a stock photo includes two components: (1) dimension and (2) resolution. These two factors play a part in the price of the stock photo selected for purchase. Normally, after a stock photo is selected, you must then decide what quality is desired. Generally, stock photos are organized in this pricing structure:

Stock Photography Quality Chart			
	Format	File Size	Usage
Low *(cheapest)*	72 dpi RGB JPEG	1MB	Web or Media
Editorial	300 dpi RGB JPEG	4MB	3.5" x 5.25" (A6)
Medium	300 dpi RGB JPEG	15MB	6" x 9" (A5)
High	300 dpi RGB JPEG	30MB	9.5" x 14" (A4)
Very High *(expensive)*	300 dpi RGB JPEG	50MB	14" x 20.5" (A3)

Some vendors, who provide on-line purchase of stock photos, will often times sell "membership" to online stock photography repositories, or purchase of CD collections include:

- Hemera.com

- ComStock.com

- iStockPhoto.com

- Corbis.com

- BananaStock.com

- Creatas.com

Carefully read through the FAQ's of stock photography vendors and also the license agreements to be sure your client's intended usage of the stock photos strictly follows the vendor's regulations. Beside Stock Photography, there exists Stock Illustrations, Stock Art, Stock Video and Clip Art. Many vendors provide colorful images, art and video to give web sites that professional edge. Such images help to replace bland buttons and boring borders with colorful and intricate images.

File Transfer Protocal (FTP) Software

Unlikely to be used by a REAL ESTATE WEBOGRAPHER™ professional; FTP software provides the ability to transfer files, web pages and images, from a personal computer (PC) to a traditional hosting company's server. When utilizing a *Webographer-friendly hosting company*, file upload via FTP is never used because they make it easy through use of *an In-browser, web development editor* as mentioned in Chapter 5. FTP software is required especially when using a web site development editor that is stand-alone, and resides on one's PC.

Commonly, a web designer will "test" a custom web site on their PC after web site creation and editing. Then web designer will upload or FTP the files to a web server. Many software programs exist to provide an interface between the PC and the server. Sometimes, FTP services are included in a professional, web development editor on your PC. If not included, there are many stand-alone FTP programs that provide for file transfer.

Such programs commonly provide a graphical interface that shows the file structure, folders and files, of the PC (origin) and the server (destination). Some examples of freeware or shareware FTP programs include:

- <u>WS FTP Pro</u>—from Ipswitch, Inc, provides for fast transfers while offering easy user navigation
- <u>CuteFTP Home 6.0</u>—from Globalscape, Inc, transfers files between home and remote computers, easy to use interface and wizard functionality for beginners.

PART SEVEN:
The Future of Technology Adoption

Chapter 16: Webographers.com

Webographers.com

The National Institute of Webographers is committed to awareness, acceptance and adoption of web-based technologies by real estate professionals and committed assistants. Simply stated, web-based technologies are an enabler to streamline many of these efforts, prior to implementing web based technologies, could not be known and appreciated by the consumer. Many of these efforts, prior to implementing web based technologies, were not made aware in full detail to the consumer.

Presently, consumers are extremely web-savvy. If one agent has a minimal-to-no web presence, consumers will go to the next agent that appears to provide all inclusive services online. A web presence makes your services visible to site-visitors, in many ways, acts as a portfolio or interactive resumé of your tech-savvy abilities.

Consumers want to be actively involved in the sale or purchase of a home. Buyers are actively performing their own market research, to supplement or cross-check an agent's findings. Buyers have many sources to self-serve their own research needs. What's important for real estate agents to understand what are buyers looking for when conducting a search online. Yes they may find your listing on Realtor.com. Does that listing include Virtual Tours? Yes they may browse your inclusive photos of the property, however, that may not be enough to hold their attention.

For sellers, the listing and advertising efforts you put forth in your web presence implies what you will do for them. If you provide a single-property website for every listing, and include virtual tours on each website, sellers then see that you could do the same for them.

Given Electronic Forms and Online Transaction Management to conduct a paperless transaction, many consumers will embrace you with "open arms"; that you enabled such technology on their behalf. Given the technology of digital signatures utilized in your transaction, who are more apprehensive, real estate professionals or consumers? One of your buyer clients may be applying for pre-approval for a home loan online; applying a "digital signature" to the loan application. Consumers are ready to embrace such technology and these savvy consumers assume that real estate professionals are doing the same.

The agent website accessed by your personal URL, is the "storefront" to you as a service provider. It implies you are readily available to site-visitors with any inquiry. Just as you may desire a timely response from a service provider you found online, so does a potential client who has encountered your web presence.

Technology Acceptance

National Institute of Webographers understands that technology can be overwhelming to the real estate professional. Sometimes, it's confusing to understand what some service providers really provide! Trust us, we reviewed many service providers of real estate applications and know what you are going through. In terms of real estate technology, the majority of the tools you need are web-based. What this book and the online training for the REAL ESTATE WEBOGRAPHER™ certification look to provide is technology acceptance by the real estate professional. What do we mean by technology acceptance?

Technology Acceptance Model
Technology Acceptance is rooted on two principles: Perceived Usefulness and Perceived Ease-of-use (e.g., Davis, Bagozzi, & Warshaw, 1989; Tannenbaum, 1990) The specific products and services showcased in this book and online training are due to *our* perceptions of their Perceived Usefulness and Perceived Ease-of-use.

During the National Institute of Webographers, LLC's two year research effort of defining "core" web-based applications that should be in your toolkit, we noted many interesting items about providers of real estate technology. Many web-based technologies may have some functionality that others do not, but they are categorized the same. Many products on the market did not fit into a "category", as their services were too broad in scope and were sometimes confusing. Many technology providers implemented product features they thought were "neat" versus what the real estate practitioner or end-user really wanted.

From those research efforts, the National Institute of Webographers has constructed the overarching competencies that we feel must be embraced by the real estate professional. Within each competency, we note web-enabled products as exemplary for a given competency as mentioned through this book.

Core Real Estate Technologies

- Agent Websites
- Single-Property Websites
- Virtual Tours
- MLS/IDX/VOW/ILD Applications
- Neighborhood Search Integration
- Comparable Market Analysis (CMA) reports
  Driven by Automated Valuation Modeled (AVM) technology
- Electronic Forms
- Online Transaction Management
- Mobile Technologies
- Virtual Assistance
  Administrators of Technology

Online Training Experience

During the online training for the REAL ESTATE WEBOGRAPHER™ certification, candidates work with "real" technologies. The Corporate Sponsors whose technologies are showcased, provide each candidate a "demo" account to work with their applications in a controlled education environment.

Why is having a demo account of each technology important in Webographers.com, online training?

> Research that extends the Technology Acceptance Model states that Motivation is another principle in end-users, like real estate professionals, accepting and using technology (e.g. Venkatesh, Speier, & Morris 2002). The online training for the REAL ESTATE WEBOGRAPHER™ certification is known as "pre-training". The training is not rigorous on the *in's* and *out's* of a product as if you had purchased, but simply gives you a taste of what is on the market, and teaches process, procedure and best practices.

How does having a user account and working with a real technology build motivation to accept and use technology?

> For some users, a demo account for each real estate technology allows a "try before you buy" a platform. For a user to fully grasp and understand a technology, they must be able to use it; versus seeing "pictures" or reading just text-based descriptions of its functionality.

Why does Webographers.com showcase 1 company per a given competency, when there are many providers in existence?

> In an environment where training, assessment and certification occur, National Institute of Webographers must "grade" candidates on work performed, where all candidates use the same technology product for a given task. This eliminates "bias" in the grading of all candidates. If two technology products were made accessible for a given task, some may say PRODUCT A is easier to use than PRODUCT B. Thus, those who scored higher in assessment using PRODUCT A than those who arbitrarily selected PRODUCT B, may have had an unfair advantage.

Features Provided to Candidates

The Webographers.com online training experience for the REAL ESTATE WEBOGRAPHER™ certification includes tools to facilitate enhance learning.

Accounts with Technology Providers
As stated, candidates are provided demo accounts on each technology platform. Each candidate is provided full use of technologies with limitations that "make sense" in a training environment. Candidates performed specific task to build understanding and comfort for the competency the product sponsors.

Training Email Account
Also, a training email address (i.e. JDoe@Webographers.com) is provided to candidates to experience the inclusive email functionality of many of the web-based technologies showcased. As stated throughout this book, many technologies provide automatic notifications to the real estate professional, account holder. Or, some applications "ignite" lead emails from web-based submission forms of site-visitors surfing the agent's web presence.

Interactive Learning Content
Multimedia-driven learning content takes the training experience well beyond still photos and paragraphs of text. Recordings of product demonstrations are

presented to add a sense of realism in understanding a product and its business value. In addition, much of the learning content is interactive, inviting active participation of the candidate.

Discounts on Product Purchasing

As stated with the Technology Acceptance Model, users are motivated and highly likely to use a product after "pre-training" with that technology. Certificants of the REAL ESTATE WEBOGRAPHER™ certification are given handsome discounts to continue product use well after training. Many of the technologies showcased in the online training allow you to keep your "work", as a good reference when using the web-based technology in the field.

Life after Certification

At Webographers.com, there is an exclusive community of REAL ESTATE WEBOGRAPHER™ certification sharing best-practices in technology use. Certificants who are logged in, are made aware of other active participants. During training, communication is limited to the candidate and course proctors. However, once certification has been obtained, any user can communicate with other users for best practice sharing.

The online training and this book, highlight core competencies that provide as general knowledge. However, the tips and best practices of practitioners using the web-based technologies in the field are expansive, creative, and innovative.

Discussion Forums

For each overarching technology competency is a respective online discussion for active question and answer sessions amongst certificants. The discussion forums not only provide knowledge-sharing amongst certificants, but provide as a lasting repository of information on best practices, and applied use of the sponsored technologies showcased in training.

Live Chat

At Webographers.com, certificants of the REAL ESTATE WEBOGRAPHER™ certification can see who else is logged in. From there, users can decide to enter in a chat session, sharing best practices in real-time.

Certification Renewal

The REAL ESTATE WEBOGRAPHER™ certification is good for up-to 2 years. During the last 2 months of certification period or 1 month after expiring, certificants can renew their certification status through a renewal course.

Road Ahead

As web-based technology that encompasses the web presence of a real estate professional evolves, so will the REAL ESTATE WEBOGRAPHER™ certification. This certification looks to stay current with real estate technology advancements, along with the rules and policies of governing bodies that govern their use.

We hope that real estate professionals will look to further accept and adopt technology that supports their business objectives. National Institute of Webographers and the REAL ESTATE WEBOGRAPHER™ professionals are here to embrace those individuals who are ready for that step. We look forward to personally meeting you at Webographers.com.

PART EIGHT:
Appendices

Appendix A: Virtual Assistant Guide per State

Appendix B: RapidListings MLS IDX Coverage

Index

APPENDIX A

Virtual Assistant Guide per State

In this appendice is a listing of what a "non-licensed" versus "licensed" individual can perform within a given state. This information is provided by Team Double-Click. National Institute of Webographers makes no warranties or claims to its accuracy.

A VA staffing company, like Team Double-Click, Inc can provide screening, oversight of work-performed, and coordination of the payment process between the real estate professional and virtual assistant. Self-managed VA's do not have such oversight.

For sales information on obtaining a virtual assistant for the individual agent, broker, or to support agency tasks, visit www.TeamDoubleClick.com, call toll free at 888.827.9129 to speak with a sales representative, or contact by email at quotes@TeamDoubleClick.com.

Disclaimer

Team Double-Click, Inc. specifically disclaims any liability, loss or risk, personal or otherwise, incurred as a consequence directly or indirectly of the use and application of any of the information herein or services offered by the company.

Team Double-Click will not be held liable for the accuracy of the information contained herein. Nor can Team Double-Click, Inc. be held responsible for violations of state laws, rules, or regulations as a result of omission or error in inclusion, or as a result of the use of the information contained herein. While great care was taken to ensure the information is/was accurate at the time of printing, it is the reader's burden and responsibility to verify accuracy.

Brokers and assistants who may refer to these "Guidelines" from time to time should be aware that it does not take very much to go from unlicensed to licensed activity. For example, it is a commonly held belief and understanding among licensees and others that participation in "negotiations" is somehow limited to the actual bargaining over terms and conditions of a sale or loan, when in fact the courts in many states have given much broader application to this term to include activity which may directly assist or aid in the negotiations or closing of a transaction.

These "Guidelines," when strictly followed, will assist licensees and their assistants to comply with the license requirements of Real Estate Law. These guidelines ARE NOT a replacement for checking the specific laws for each state.

This listing is not intended as a replacement for looking up and knowing the exact laws with regards to unlicensed real estate assistants. It is intended, rather, as a quick reference guide for your convenience.

Sources

In all instances the information contained herein was acquired directly from either the State's Real Estate Association or the State's Real Estate Commission.

General notes about the use of unlicensed assistants, virtual or otherwise

The unlicensed assistant MUST identify him or herself as unlicensed

In some states stepping beyond the bounds of an unlicensed assistant is a FELONY charge and can carry fines of over $10,000 AND jail time for the assistant AND the broker or agent allowing/requesting an unlicensed assistant to perform the duties of a licensee.

In all states, an unlicensed assistant MAY NOT perform any duties for which the State requires a license.

The training our virtual assistants receive through Team Double-Click, Inc. (Team Double-Click's TCRE—Team Contractor Real Estate Elite) is NOT considered a real estate license. It is private certifications helping these independent contractors understand how to work as an assistant for a real estate agent or broker.

ALABAMA   AL

Alabama Real Estate Commission: http://www.arec.state.al.us

An unlicensed assistant May:

1. Answer the telephone, forward calls, take messages, and make appointments for licensees.
2. Send listing information to a multiple listing service, filling out the necessary forms.
3. Deliver information and forms to a mortgage company and closing attorney or agent as part of the preparation for closing.
4. Make and deliver copies of any public record.
5. Get keys from a client/owner and have keys made.
6. Write and place advertising in newspaper and other forms of publication.
7. Receive and deposit funds to be held in trust for others including earnest money, security deposits, and rental payments.
8. Type forms.
9. Perform company bookkeeping.
10. Place signs on property.
11. Arrange for and oversee repairs.
12. Make rental collection calls to tenants.
13. Answer questions about a property as long as the answers are available in some pre-printed form.
14. Give a key to a prospect.

An unlicensed assistant May Not:

1. Prepare or discuss a listing or property management agreement with an owner.
2. Show any property or be at an open house for any purpose.
3. Drive or accompany a prospect to a property.
4. Negotiate or discuss the terms of a sale or rental.
5. Procure prospects by door to door visits or canvass type telephone calls.
6. Prepare or have a prospect sign an offer to purchase or lease.
7. Present an offer to an owner.

ALASKA   AK

Alaska Real Estate Commission: http://www.dced.state.ak.us/occ/prec.htm

An unlicensed assistant May Not:

1. A licensee may not authorize an unlicensed assistant to perform any duties for which a license is required, including:
2. Discussing a listing or property management agreement with an owner or with licensees;
3. Showing any property available for sale or rental;
4. Negotiating or discussing the terms of a sale or rental;
5. Having a prospective buyer or lessee sign an offer to purchase or lease;
6. Presenting an offer to a seller;
7. Making prospecting calls or visits; and
8. Reading prepared information in response to inquiries about properties.

ARIZONA   AZ

Arizona Real Estate Commission: http://www.re.state.az.us

An unlicensed assistant May:

1. Perform telephone duties, to include calls to:
   a. collect demographic information
   b. solicit interest in engaging the services of a licensee or brokerage
   c. set or confirm appointments (with no other discussion) for:
      i. A licensee to list or show property
      ii. A buyer with a loan officer
      iii. A property inspector to inspect a home
      iv. A repair/maintenance person to perform repairs/maintenance
      v. An appraiser to appraise property
2. Mortgage and/or title companies to track the status of a file, check daily interest rates and points, whether buyer has been qualified, confirm closing appointment for licensee, and so forth
3. Assist a licensee at an open house
4. Unlock a home for a licensee so that licensee can show a buyer the property or preview the property (no discussion about the property).
5. Deliver documents (as a mail or delivery service only)

An unlicensed assistant May Not:

1. Hold/host an open house without an agent being present
2. Perform a walk-through inspection
3. Answer questions relating to a transactional document
4. Give instructions to inspectors, appraisers or maintenance/repair people. Because these instructions are part of the licensee's regular duties and there is a direct relationship to the (potential) transaction, a license is required in order to give instructions to inspectors appraisers or repair/maintenance people

ARKANSAS   AR

Arkansas Real Estate Commission: http://www.state.ar.us/arec/arecweb.html

Specific information about unlicensed real estate assistants: http://www.state.ar.us/arec/faq.htm#UNLICENSED

No other information is available

CALIFORNIA   CA

California Real Estate Advisory Commission: http://www.dre.ca.gov/reac.htm

Guidelines For Unlicensed Assistants *Preamble*

The designated officer of a corporation is explicitly responsible for the supervision and control of the activities conducted on behalf of a corporate broker by its officers and employees as necessary to secure full compliance with the Real Estate Law, including but not limited to the supervision of salespersons licensed to the corporation in the performance of acts for which a real estate license is required. It is inherent with respect to individuals engaging in business as a real estate broker that they are also similarly charged with the responsibility to supervise and control all activities performed by their employees and agents in their name during the course of a transaction for which a real estate license is required, whether or not the activities performed require a real estate license.

To assist brokers and designated broker/officers to properly carry out their duty to supervise and control activities conducted on their behalf during the course of a licensed transaction, it is important for the broker to know and identify those activities which do and do not require a real estate license. This knowledge assists the broker to use licensed persons when required, and to extend and provide the necessary quantum of supervision and control over licensed and non-licensed activities as required by law and good business practices.

Identifying licensed activities has become difficult for many brokers as brokerage practices have changed and evolved in response to new laws, the need for new efficiencies in response to consumer demands, and new technology. The following is a guideline, and nothing more, of defined activities which generally do not come within the term "real estate broker," when performed with the broker's knowledge and consent. Broker knowledge and consent is a prerequisite to the performance of these unlicensed activities, since without these elements there can be no reasonable assurance that the activities performed will be limited as set forth below.

An unlicensed assistant May (note stipulations to such):

1. Cold Call
    a. Making telephone calls to canvass for interest in using the services of a real estate broker. Should the person answering the call indicate an interest in using the services of a broker, or if there is an interest in ascertaining the kind of services a broker

can provide, the person answering shall be referred to a licensee, or an appointment may be scheduled to enable him or her to meet with a broker or an associate licensee** (licensee***). At no time may the caller attempt to induce the person being called to use a broker's services. The canvassing may only be used to develop general information about the interest of the person answering and may not be used, designed or structured for solicitation purposes with respect to a specific property, transaction or product. (The term "solicitation" as used herein should be given its broadest interpretation.)

2. Open Houses
    a. With the principal's consent, assisting licensees at an open house intended for the public by placing signs, greeting the public, providing factual information from or handing out preprinted materials prepared by or reviewed and approved for use by the licensee, or arranging appointments with the licensee. During the holding of an open house, only a licensee may engage in the following: show or exhibit the property, discuss terms and conditions of a possible sale, discuss other features of the property, such as its location, neighborhood or schools, or engage in any other conduct which is used, designed or structured for solicitation purposes with respect to the property.

3. Comparative Market Analysis
    a. Making, conducting or preparing a comparative market analysis subject to the approval of and for use by the licensee.

4. Communicating With the Public
    a. Providing factual information to others from writings prepared by the licensee. A non-licensee may not communicate with the public in a manner which is used, designed or structured for solicitation purposes with respect to a specific property, transaction or product.

5. Arranging Appointments
    a. Making or scheduling appointments for licensees to meet with a principal or party to the transaction. As directed by the licensee to whom the broker has delegated such authority, arranging for and ordering reports and services from a third party in connection with the transaction, or for the provision of services in connection with the transaction, such as a pest control inspection and report, a roof inspection and report, a title inspection and/or a preliminary report, an appraisal and report, a credit check and report, or repair or other work to be performed to the property as a part of the sale.

6. Access to Property
    a. With the principal's consent, being present to let into the property a person who is either to inspect a portion or all of the property for the purpose of preparing a report or issuing a clearance, or who is to perform repair work or other work to the property in connection with the transaction. Information about the real property

which is needed by the person making the inspection for the purpose of completing his or her report must be provided by the broker or associate licensee, unless it comes from a data sheet prepared by the broker, associate licensee or principal, and that fact is made clear to the person requesting the information.

7. Advertising

   a. Preparing and designing advertising relating to the transaction for which the broker was employed, if the advertising is reviewed and approved by the broker or associate licensee prior to its publication.

8. Preparation of Documents

   a. Preparing and completing documents and instruments under the supervision and direction of the licensee if the final documents or instruments will be or have been reviewed or approved by the licensee prior to the documents or instruments being presented, given or delivered to a principal or party to the transaction.

9. Delivery and Signing Documents

   a. Mailing, delivering, picking up, or arranging the mailing, delivery, or picking up of documents or instruments related to the transaction, including obtaining signatures to the documents or instruments from principals, parties or service providers in connection with the transaction. Such activity shall not include a discussion of the content, relevance, importance or significance of the document, or instrument or any portion thereof, with a principal or party to the transaction.

10. Trust Funds

   a. Accepting, accounting for or providing a receipt for trust funds received from a principal or a party to the transaction.

11. Communicating With Principals, etc.

   a. Communicating with a principal, party or service provider in connection with a transaction about when reports or other information needed concerning any aspect of the transaction will be delivered, or when certain services will be performed or completed, or if the services have been completed.

12. Document Review

   a. Reviewing, as instructed by the licensee, transaction documentation for completeness or compliance, providing the final determination as to completeness or compliance is made by the broker or associate licensee.

   b. Reviewing transaction documentation for the purpose of making recommendations to the broker on a course of action with respect to the transaction.

■ These "Guidelines," when strictly followed, will assist licensees and their employees to comply with the license requirements of the Real Estate Law. They present specific scenarios which allow brokers to organize their business practices in a manner that will contribute to compliance with the Real Estate Law. As such, they were drafted to serve the interests of both licensees and the public they serve. Nothing in them is intended to limit, add to or supersede any provision of law relating to the duties and obligations of real estate licensees, the consequences of violations of law or licensing requirements.

- Licensees should take heed that because of the limiting nature of guidelines, as opposed to a statute or regulation, that they will not bind or obligate, nor are they intended to bind and obligate courts or others to follow or adhere to their provisions in civil proceedings or litigation involving conduct for which a real estate license may or may not be required.
- Brokers and others who may refer to these "Guidelines" from time to time should be aware that it does not take very much to go from unlicensed to licensed activity. For example, it is a commonly held belief and understanding among licensees and others that participation in "negotiations" is somehow limited to the actual bargaining over terms and conditions of a sale or loan, when in fact the courts in this state have given much broader application to this term to include activity which may directly assist or aid in the negotiations or closing of a transaction.
- The term "associate licensee" means and refers to either a salesperson employed by the listing or selling broker in the transaction, or a broker who has entered into a written contract with a broker to act as the broker's agent in transactions requiring a real estate license.
- Hereafter, the term "licensee" means "broker" or "associate licensee."

COLORADO   CO

Colorado Division of Real Estate:
http://www.dora.state.co.us/real-estate/index.htm

Specific information about unlicensed real estate assistants:
http://www.dora.state.co.us/real-estate/manual/Manual2005/chap3.pdf see statement 20

CP-20 Commission Position Statement On Personal Assistants (Adopted August 2, 2001)

The use of personal assistants has grown considerably in recent years. Personal assistants are generally thought of as unlicensed persons performing various functions as employees (including clerical support) or independent contractors of a real estate broker within the framework of a real estate transaction. The Commission recognizes the growth in the utilization of such assistants. Inquiries generally fit into two categories: (1) whether the activity performed is one which requires a license, and (2) what are the supervisory responsibilities of an employing broker.

The license law prohibits unlicensed persons from negotiating, listing or selling real property. Therefore, foremost to the use of personal assistants is careful restriction of their activities so as to avoid illegal brokerage practice. Personal assistants may complete forms prepared and as directed by licensees but should never independently draft legal documents such as listing and sales contracts, nor should they offer opinions, advice or interpretations, in addition, they should not distribute information on listed properties other than that prepared by a broker.

An unlicensed assistant May

1. Perform clerical duties for a broker which may include the gathering of information for a listing;
2. Provide access to a property and hand out preprinted, objective information, so long as no negotiating, offering, selling or contracting is involved:
3. Distribute preprinted, objective information at an open house, so long as no negotiating, offering, selling or contracting is involved;
4. Distribute information on listed properties when such information is prepared by a broker;
5. Deliver paperwork to other brokers;
6. Deliver paperwork to sellers or purchasers, if such paperwork has already been reviewed by a broker;
7. Deliver paperwork requiring signatures in regard to financing documents that are prepared by lending institutions; and
8. Prepare market analyses for sellers or buyers on behalf of a broker, but disclosure of the name of the preparer must be given, and it must be submitted by the broker.

Employing brokers need to be especially aware of their supervisory duties under the license law. Supervisory duties apply whether the assistant is an employee or independent contractor.

An employing broker should have a written office policy explaining the duties, responsibilities and limitations on the use of personal assistants. This policy should be reviewed by and explained to all employees.

Licensees should not share commissions with unlicensed assistants. Although this may not technically be a violation of the licensing act if the activity is not one which requires a license, the temptation to "cross over" into the area of negotiating and other prohibited practices is greatly increased where compensation is based on the success of the transaction.

If brokers develop adequate policies for the use of assistants and routine procedures for monitoring their activities, the assistant can serve as a valuable tool in the success of the transaction. As with any other activity involving the delegation of an act to another, the freedom and convenience afforded the broker in allowing the use of assistants carries with it certain responsibilities for that person's actions.

CONNECTICUT   CT

Connecticut Real Estate Commission: http://www.ct.gov/dcp/cwp/view.asp?a=1624&Q=276076

Specific information on unlicensed assistants:
http://www.business.uconn.edu/RealEstate/Library/Oct04RELicensingLaws.pdf see page 48

The Connecticut Real Estate Commission prohibits unlicenced persons from negotiating, listing, selling, buying, or renting real property for another for a fee. It is, therefore, important for employing brokers and other licensees using such persons to carefully restrict the activities of such persons so that allegations of wrongdoing under Connecticut General Statutes or State Regulations can be avoided.

Licensees should not share commissions with unlicenced persons acting as assistants, clerical staff, closing secretaries, etc. The temptation for such unlicenced persons, in such situations, to go beyond what they can do and negotiate or take part in other prohibited activities is greatly increased when their compensation is based on the successful completion of the sale.

In order to provide guidance to licensees with regard to which activities related to a real estate transaction unlicenced persons can and cannot perform, the commission establishes the following Policy:

An unlicensed assistant May

Activities which *may* be performed by unlicenced persons who, for example, act as personal assistants, clerical support staff, closing secretaries, etc., include, but are not necessarily limited to:

1. Host open houses, kiosks, home show booths or fairs, or hand out materials at such functions.
2. Show property.
3. Answer any questions from consumers on listing, title, financing, closing, etc.
4. Contact cooperative brokers, whether in person or otherwise, regarding any negotiations or open transactions.
5. Discuss or explain a contract, offer to purchase, agreement, listing, or other real estate document with anyone outside the firm.
6. Be paid on the basis of commission, or any amount based on listings, sales, etc.
7. Negotiate or agree to any commission, commission split or referral fee on behalf of a licensee.
8. Place calls that would require a license such as cold calls, solicit listings, contacting expired listings or for sale by owners, or extending invitations to open houses.
9. Attend inspections or pre-closing walk-through unless accompanied by licensee.
10. The unlicensed assistant is not a decision maker; rather, shall take all direction from supervising licensee.

An unlicensed assistant May Not:

Activities which *cannot* be performed by unlicenced persons who, for example, act as personal assistants, clerical support staff, closing secretaries, etc., include but are not necessarily limited to:

1. Host open houses, kiosks, home show booths or fairs, or hand out materials at such functions.
2. Show property.
3. Answer any questions from consumers on listing, title, financing, closing, etc.
4. Contact cooperative brokers, whether in person or otherwise, regarding any negotiations or open transactions.

5. Discuss or explain a contract, offer to purchase, agreement, listing, or other real estate document with anyone outside the firm.

6. Be paid on the basis of commission, or any amount based on listings, sales, etc.

7. Negotiate or agree to any commission, commission split or referral fee on behalf of a licensee.

8. Place calls that would require a license such as cold calls, solicit listings, contacting expired listings or for sale by owners, or extending invitations to open houses.

9. Attend inspections or pre-closing walk-through unless accompanied by licensee.

10. The unlicenced assistant is not a decision maker; rather, shall take all direction from supervising licensee.

Employing brokers, whether they are employing unlicenced persons or whether licensees under their supervision are using unlicenced persons as personal assistants or the like, are responsible for assuring that such unlicenced persons are not involved in activities which require a license and/or activities which violate this policy. Brokers should establish guidelines for the use of unlicenced persons and procedures for monitoring *State of Connecticut-October 2003* their activities. It is the responsibility of the designated broker to assure at unlicenced persons, either directly employed or contracted, or employed or contracted by licensees under his or her supervision, are not acting improperly.

DELAWARE   DE

Delaware Real Estate Commission: http://www.dpr.delaware.gov/boards/realestate

See "License Law" at http://www.dpr.delaware.gov/boards/realestate/index.shtml 24 Del. C. § 2901, a license is required.

Definition of broker below. Unlicensed assistants may not perform these activities:

1. "Real estate broker" means any person who, for a compensation or valuable consideration, sells or offers for sale, buys or offers to buy, or negotiates a purchase, sale or exchange of real estate or who leases or offers to lease or rents or offers for rent any real estate or the improvements thereon for others, as a whole or partial vocation, but shall not include an auctioneer as defined in § 2301(a)(3) of Title 30.

2. "Real estate salesperson" means any person who, for a compensation or valuable consideration, is employed, either directly or indirectly by a real estate broker, to sell or offer to sell, or to buy or to offer to buy, or to negotiate the purchase or sale or exchange of real estate, or to lease or rent or offer for rent any real estate, or to negotiate leases thereof or of the improvements thereon, as a whole or partial vocation, but shall not include an auctioneer as defined in § 2301(a)(3) of Title 30.

DISTRICT OF COLUMBIA   DC

District of Columbia Board of Real Estate: http://app.dcra.dc.gov/information/build_pla/occupational/real_estate/index.shtm

No other information is available at time of printing

FLORIDA   FL

Florida Division of Real Estate: http://www.state.fl.us/dbpr/re/index.shtml

An unlicensed assistant May:

1. Answer the phone and forward calls.
2. Fill out and submit listings and changes to any multiple listing service.
3. Follow up on loan commitments after a contract has been negotiated and generally secure status reports on the progress of the loan.
4. Assemble documents for closing.
5. Secure public information from courthouses, utility districts, etc.
6. Have keys made for company listings.
7. Write advertisements for approval of licensee and supervising broker, and place classified advertising.
8. Receive, record and deposit earnest money, security deposits and advance rents.
9. Type contract forms and Supreme Court-approved leases (those leases approved by the Supreme Court for use by non-lawyers), for approval by licensee and supervising broker.
10. Monitor licenses and personnel files.
11. Compute commission checks.
12. Place signs on property.
13. Order items of repair as directed by the licensee.
14. Prepare fliers and promotional information for approval by licensee and supervising broker.
15. Deliver documents and pick up keys.
16. Place routine telephone calls regarding late rent payments.
17. Schedule appointments for licensee to show listed property.
18. Be present at open houses to provide security, hand out materials (brochures) and respond to questions that may be answered with objective responses gleaned from pre-printed objective information.
19. Gather information for a comparative market analysis (CMA).
20. Gather information for an appraisal.
21. Hand out objective, written information on a listing or rental.
22. Drive a customer or client to a listing or rental (however, an unlicensed assistant may not provide access to a listed property for sale or lease.)
23. Give a key to a prospect.

An unlicensed assistant May Not:

Further, FREC ruled that an unlicensed individual may not negotiate or agree to any commission split or referral fee on behalf of a licensee (FREC reversed an earlier ruling that this was permissible).

GEORGIA   GA

Georgia Real Estate Commission: http://www.grec.state.ga.us

Links to specific information about unlicensed assistants:
http://www.grec.state.ga.us/grec/legal/title_43/43_40_25.html
http://www.grec.state.ga.us/grec/legal/title_43/43_40_29.html

An unlicensed assistant May Not

1. Discuss, negotiate, or explain a contract, listing, buyer agency, lease, agreement, or other real estate document;
2. Vary or deviate from the rental price or other terms and conditions previously established by the owner or licensee when supplying relevant information concerning the rental of property;
3. Approve applications or leases or settle or arrange the terms and conditions of a lease;
4. Indicate to the public that the unlicensed individual is in a position of authority which has the managerial responsibility of the rental property;
5. Conduct or host an open house or manage an on-site sales office;
6. Show real property;
7. Answer questions regarding company listings, title, financing, and closing issues, except for information that is otherwise publicly available;
8. Discuss, negotiate, or explain a contract, listing, buyer agency, lease, agreement, or other real estate document;
9. Be paid solely on the basis of real estate activity including, but not limited to, a percentage of commission or any amount based on the listing or sales compensation or commission;
10. Negotiate or agree to compensation or commission including, but not limited to, commission splits, management fees, or referral fees on behalf of a licensee;
11. Engage in an activity requiring a real estate license as required and defined by this chapter.

HAWAII   HI

Hawaii Department of Commerce and Consumer Affairs—Real Estate Branch:
http://www.hawaii.gov/dcca/areas/real

Specific information on real estate licensing:
http://www.hawaii.gov/dcca/areas/real/main/hrs click HRS Chapter 467

No other information is available at time of printing

IDAHO   ID

Idaho Real Estate Commission: http://www.idahorealestatecommission.com

An unlicensed assistant May:

1. Perform clerical duties for an employing broker or broker associate which may include the gathering of information for a listing;
2. Provide access to a property other than showings to potential buyers and hand out reprinted, objective information, so long as no negotiating, offering, selling or contracting is involved;
3. Distribute preprinted, objective information at an open house, so long as no negotiating, offering, selling or contracting is involved;
4. Distribute information on listed properties when such information is prepared by the broker or broker associate;
5. Deliver paperwork to other brokers;
6. Deliver paperwork to sellers or purchasers, if such paperwork has already been reviewed by a broker;
7. Deliver paperwork requiring signatures in regard to financing documents that are prepared by lending institutions; and
8. Prepare market analyses for sellers or buyers on behalf of a broker, but disclosure of the name of the preparer must be given, and it must be submitted by the broker.

ILLINOIS   IL

Illinois Division of Banks and Real Estate: www.obre.state.il.us/realest/realmain.htm

Specific information on unlicensed assistants:
http://www.ilga.gov/commission/jcar/admincode/068/068014500D01650R.html

Licensees under the Act may employ, or otherwise utilize the services of, unlicensed assistants to assist them with administrative, clerical, or personal activities for which a license under the Act is not required.

An unlicensed assistant, on behalf of and under the direction of a licensee, may engage in the following administrative, clerical, or personal activities without being in violation of the licensing requirements of the Act. The following list is intended to be illustrative and declarative of the Act and is not intended to increase or decrease the scope of activities for which a license is required under the Act.

An unlicensed assistant May

1. Answer the telephone, take messages, and forward calls to a licensee;
2. Submit listings and changes to a multiple listing service;
3. Follow up on a transaction after a contract has been signed;
4. Assemble documents for a closing;
5. Secure public information from a courthouse, sewer district, water district, or other repository of public information;
6. Have keys made for a company listing;
7. Draft advertising copy and promotional materials for approval by a licensee;
8. Place advertising;
9. Record and deposit earnest money, security deposits, and rents;
10. Complete contract forms with business and factual information at the direction of and with approval by a licensee;
11. Monitor licenses and personnel files;
12. Compute commission checks and perform bookkeeping activities;
13. Place signs on property;
14. Order items of routine repair as directed by a licensee;
15. Prepare and distribute flyers and promotional information under the direction of and with approval by a licensee;
16. Act as a courier to deliver documents, pick up keys, etc.;
17. Place routine telephone calls on late rent payments;
18. Schedule appointments for the licensee (this does not include making phone calls, tele-marketing, or performing other activities to solicit business on behalf of the licensee);
19. Respond to questions by quoting directly from published information;
20. Sit at a property for a broker tour which is not open to the public;
21. Gather feedback on showings;
22. Perform maintenance, engineering, operations or other building trades work and answer questions about such work;
23. Provide security;
24. Provide concierge services and other similar amenities to existing tenants;
25. Manage or supervise maintenance, engineering, operations, building trades and security; and
26. Perform other administrative, clerical, and personal activities for which a license under the Act is not required.

An unlicensed assistant May Not:

An unlicensed assistant of a licensee **may not** perform the following activities for which a license under the Act is required. The following list is intended to be illustrative and declarative of the Act and is not intended to increase or decrease the scope of activities for which a license is required under the Act. An unlicensed assistant of a licensee may not:

1. host open houses, kiosks, or home show booths or fairs;
2. show property;
3. interpret information on listings, titles, financing, contracts, closings, or other information relating to a transaction;
4. explain or interpret a contract, listing, lease agreement, or other real estate document with anyone outside the licensee's company;
5. negotiate or agree to any commission, commission split, management fee, or referral fee on behalf of a licensee; or
6. perform any other activity for which a license under the Act is required.
7. Any licensee who employs an unlicensed assistant shall be responsible for the actions of the unlicensed assistant taken while under the supervision of or at the direction of the licensee.
8. Any licensee who is responsible for the actions of an unlicensed assistant by statute, regulation, contract, or office policy and who permits, aids, assists, or allows an unlicensed assistant to perform any activity for which a license under the Act is required shall be in violation of the Act.
9. Stenographic, clerical, maintenance, engineering, building trades, security, or office personnel not directly engaged in the practice of real estate brokerage as defined in Section 1-10 of the Act are not required to be licensed.
10. A licensee is prohibited from acting as an unlicensed assistant for any licensee other than his or her sponsoring broker or a licensee sponsored by the sponsoring broker.

INDIANA   IN

Indiana Real Estate Commission: http://www.in.gov/pla/bandc/estate

An unlicensed assistant May

1. Answer the phone, forward calls to a licensee, give out addresses, directions and price lists.
2. Submit listings and changes to a multiple listing service.
3. Follow up on loan commitments after a contract has been negotiated.
4. Assemble documents for the closings.
5. Obtain documents and information from the courthouse, utilities offices, title companies, and others.
6. Have keys made for company listings.
7. Write ads for approval of the licensee and supervising broker and place advertising (promotional information, newspaper ads, etc.)
8. Record and deposit earnest money, security deposits, and advance rents.
9. Type (not create) contract forms for approval by the licensee and supervising broker.
10. Monitor licenses and personnel files
11. Compute commission checks.
12. Place signs on property.

13. Order items of routine repair as directed by the licensee.
14. Prepare flyers and promotional information for approval by the licensee and supervising broker.
15. Act as courier service to deliver documents, pick up keys, etc.
16. Schedule appointments for the licensee to show listed property

An unlicensed assistant May Not

1. Prepare promotional materials or ads without the review and approval of licensee and supervising broker.
2. Show property
3. Answer general questions on listings, title, financing, closings, or otherwise negotiate.
4. Discuss or explain a contract, listing, lease, agreement, or other real estate document with anyone outside the firm.
5. Negotiate or agree to any commission, commission split, management fee or referral fee on behalf of a licensee.
6. Answer general questions from a listing sheet including explaining, promoting, or negotiating with regard to listed property.
7. Conduct an open house.
8. Conduct telemarketing of telephone canvassing to seek listings.

IOWA   IA

Iowa Professional Licensing Division—Real Estate Sales Persons and Brokers Office: http://www.state.ia.us/government/com/prof/sales/home.html

193E–7.13(543B) Support personnel for licensees; permitted and prohibited activities.

Whenever a licensee affiliated with a broker engages support personnel to assist the affiliated licensee in the activities of the real estate brokerage business, both the firm or sponsoring broker and the affiliated licensee are responsible for supervising the acts or activities of the personal assistant; however, the affiliated licensee shall have the primary responsibility for supervision. Unless the support person holds a real estate license, the support person may not perform any activities, duties, or tasks of a real estate licensee as identified in Iowa Code sections 543B.3 and 543B.6 and may perform only ministerial duties that do not require discretion or the exercise of the licensee's own judgment. Personal assistants shall be considered support personnel.

7.13(1) Individuals actively licensed with one firm or broker may not work as support personnel for a licensee affiliated with another firm or broker. Individuals with an inactive status license may work as support personnel for a licensee, but shall not participate in any activity that requires a real estate license.

7.13(2) Any real estate brokerage firm or broker that allows an affiliated licensee to employ, or engage under an independent contractor agreement, support personnel to assist the affiliated licensee in carrying out brokerage activities must comply with the following:

    a.   Implement a written company policy authorizing the use of support personnel by licensees;

    b.   Specify in the written company policy, which may incorporate the duties listed in 7.13(4), any duties that the support personnel may perform on behalf of the affiliated licensee;

    c.   Ensure that the affiliated licensee and the support personnel receive copies of the duties that support personnel may perform.

7.13(3) Broker supervision and improper use of license and office. While individual and designated brokers shall be responsible for supervising the real estate related activities of all support personnel, an affiliated licensee employing a personal assistant shall have the primary responsibility for supervision of that personal assistant. A broker shall not be held responsible for inadequate supervision if:

    a.   The unlicensed person violated a provision of Iowa Code chapter 543B or of commission rules that is in conflict with the supervising broker's specific written policies or instructions;

    b.   Reasonable procedures have been established to verify that adequate supervision was being provided;

    c.   The broker, upon hearing of the violation, attempted to prevent or mitigate the damage;

    d.   The broker did not participate in the violation; and

    e.   The broker did not attempt to avoid learning of the violation.

7.13(4) In order to provide reasonable assistance to licensees and their support personnel, but without defining every permitted activity, the commission has identified certain tasks that unlicensed support personnel under the direct supervision of a licensee affiliated with a firm or broker may and may not perform.

a. **Permitted** activities include, but are not limited to, the following:

    1.   Answer the telephone, provide information about a listing to other licensees, and forward calls from the public to a licensee;

    2.   Submit data on listings to a multiple listing service;

    3.   Check on the status of loan commitments after a contract has been negotiated;

    4.   Assemble documents for closings;

    5.   Secure documents that are public information from the courthouse and other sources available to the public;

    6.   Have keys made for company listings;

    7.   Write advertisements and promotional materials for the approval of the licensee and supervising broker;

    8.   Place advertisements in magazines, newspapers, and other media as directed by the supervising broker;

9. Record and deposit earnest money, security deposits, and advance rents, and perform other bookkeeping duties;
10. Type contract forms as directed by the licensee or the supervising broker;
11. Monitor personnel files;
12. Compute commission checks;
13. Place signs on property;
14. Order items of routine repair as directed by a licensee;
15. Act as courier for such purposes as delivering documents or picking up keys. The licensee remains responsible for ensuring delivery of all executed documents required by Iowa law and commission rules;
16. Schedule appointments with the seller or the seller's agent in order for a licensee to show a listed property;
17. Arrange dates and times for inspections;
18. Arrange dates and times for the mortgage application, the preclosing walk–through, and the closing;
19. Schedule an open house;
20. Perform physical maintenance on a property; or
21. Accompany a licensee to an open house or a showing and perform the following functions as a host or hostess:
22. Open the door and greet prospects as they arrive;
23. Hand out or distribute prepared printed material;
24. Have prospects sign a register or guest book to record names, addresses and telephone numbers;
25. Accompany prospects through the home for security purposes and not answer any questions pertaining to the material aspects of the house or its price and terms.

b. **Prohibited** activities include, but are not limited to, the following:

1. Making cold calls by telephone or in person or otherwise contacting the public for the purpose of securing prospects for listings, leasing, sale, exchanges, or property management;
2. Independently hosting open houses, kiosks, home show booths, or fairs;
3. Preparing promotion materials or advertisements without the review and approval of licensee and supervising broker;
4. Showing property independently;
5. Answering any questions on title, financing, or closings (other than time and place);
6. Answering any questions regarding a listing except for information on price and amenities expressly provided in writing by the licensee;
7. Discussing or explaining a contract, listing, lease, agreement, or other real estate document with anyone outside the firm;
8. Negotiating or agreeing to any commission, commission split, management fee, or referral fee on behalf of a licensee;

9. Discussing with the owner of real property the terms and conditions of the real property offered for sale or lease;

10. Collecting or holding deposit moneys, rent, other moneys or anything of value received from the owner of real property or from a prospective buyer or tenant;

11. Providing owners of real property or prospective buyers or tenants with any advice, recommendations or suggestions as to the sale, purchase, exchange, rental, or leasing of real property that is listed, to be listed, or currently available for sale or lease; or

12. Holding one's self out in any manner, orally or in writing, as being licensed or affiliated with a particular firm or real estate broker as a licensee.

193E—7.14(543B) Information provided by nonlicensed support personnel restricted. Nonlicensed support personnel may, on behalf of the employer licensee, provide information concerning the sale, exchange, purchase, rental, lease, or advertising of real estate only to another licensee. Support personnel shall provide information only to another licensee that has been provided to the personnel by the employer licensee either verbally or in writing.

KANSAS   KS

Kansas Real Estate Commission: http://www.accesskansas.org/krec

Unlicensed assistants May:

1. Answer the phone and forward calls to a licensee.

2. Submit listings and changes to a multiple listing service if the listings or changes are based upon data compiled and provided by a licensed broker or salesperson.

3. Follow up on loan commitments after a contract has been negotiated.

4. Assist a broker or salesperson in assembling documents for closings.

5. Secure documents (public information) from courthouse, sewer district, water district, etc.

6. Have keys made for company listings. Place "for sale" signs on property at the direction of a broker or salesperson with the firm.

7. Write ads and prepare flyers and promotional information for approval by licensee and supervising broker and place advertising.

8. Type offers, contracts and leases from drafts prepared by a broker or salesperson with the firm.

9. Monitor licenses and personnel files. Compute commission checks.

10. Maintain trust account records under the supervision of the broker. (The broker remains responsible for compliance with the license act and regulations.)

11. Order items of routine repair as directed by a licensee with the firm.

12. Act as courier service to deliver or pick up documents, keys, etc.

13. Measure house under supervision of licensee. (The licensee and supervising broker remain responsible for accuracy of measurements.)

14. Schedule appointments for a licensee to show listed property.

15. Furnish information from listing sheets by telephone to other real estate offices. Such person may not explain or interpret information on the listing sheets.
16. Host open houses for licensees if serving strictly as a monitoring host. Greet prospective buyers and hand them printed information prepared by the builder, owner or licensee. MAY NOT explain or interpret information, discuss or make representations about the terms of sale, the home or property, or solicit new listings or new clients. ALL questions must be referred to the owner or a licensee.

NOTE The commission does not recommend that unlicensed individuals host open houses; however, within very narrow restrictions, the activity is permissible under the license act and is therefore included in this list. Brokers who choose to allow an unlicensed person to host an open house are strongly urged to closely monitor such activity. If the unlicensed person goes beyond what is permissible, the broker remains responsible.

Unlicensed assistants May Not

1. Answer questions concerning properties listed with the firm, except to confirm that the property is listed and to identify the listing broker or salesperson.
2. Show property and discuss anything related to the property or related to its purchase.
3. Discuss or explain a contract, listing, lease agreement or other real estate document with anyone outside the firm.
4. Negotiate or agree to any commission, commission split, or referral fee on behalf of a licensee.
5. A licensee may not, as a personal assistant for another licensee or as a secretary/employee, perform any activity which requires a license while licensed with another firm.

KENTUCKY   KY

Kentucky Real Estate Commission: http://krec.ky.gov

Specific information on unlicensed assistants: http://krec.ky.gov/pdf/KRECnewsfall03.pdf

An unlicensed assistant May:

1. Hold an open house and distribute literature so long as the seller agrees in writing.
2. Copy a key of a piece of property at the direction of the supervising licensee.
3. Open the door of a property with the consent of the owner.
4. Answer whether a piece of property is listed with the company.
5. Answer whether the property is under contract.
6. Answer whether the property has closed.
7. Give out the listing price of the property.
8. In writing give out the square footage, number of rooms of the property, etc.

9. Be pictured in advertisements as long as it is clear that the unlicensed assistant is not a real estate licensee.
10. Contact consumers for the purpose of setting up appointments.
11. Receive confidential information about a piece of property as long as he or she only discloses it to the supervising licensee.

An unlicensed assistant May NOT

1. Show property or answer any questions about the property.
2. Negotiate the terms of a contract.
3. Complete offers or contracts.
4. Disclose confidential or non-public information about a property that is not available to the general public.
5. Attend a real estate closing without a supervising licensee.
6. Access trade organization information if the supervising licensee is not a member of that trade group
7. Write or place advertisements without supervising licensee's review.
8. Express material opinions about a particular real estate transaction to anyone other than the supervising licensee.
9. Interpret contractual language for others.
10. Represent that he/she is a real estate licensee.

LOUISIANA   LA

Louisiana Real Estate Commission: http://www.lrec.state.la.us

Unlicensed Assistants May:

1. Answer the phone and forward calls to licensee
2. Submit listings and changes to a multiple listing service
3. Follow up on loan commitments after a contract has been negotiated
4. Place signs on listed property
5. Order items of routine repair as directed by licensee
6. Prepare flyers and promotional information for approval by licensee and supervising broker
7. Type contract forms as directed by licensee and supervising broker
8. Act as courier service to deliver documents, pick up keys, etc.
9. Schedule appointments for licensee to show listed property
10. Secure public information documents from courthouse, sewer district, water district, etc.
11. Have keys made for company listings
12. Write ads as directed by licensee and supervising broker and place advertising (promotional information, newspaper ads, etc.)

<u>*Unlicensed Assistants May Not:*</u>

1. Host an open house
2. Prepare promotional material or ads without the review and approval of licensee and supervising broker
3. Show property listed for sale
4. Answer any questions on listing
5. Discuss or explain a contract, listing, or other real estate document with anyone outside the firm
6. Be paid on the basis of real estate activity, such as a percentage of commission, or any amount based on listings, sales, etc.
7. Negotiate or agree to any commission, commission split, management fee or referral fee on behalf of a licensee

MAINE   ME

Maine Real Estate Commission: http://www.state.me.us/pfr/olr/categories/cat38.htm

Specific information about licensees:
http://janus.state.me.us/legis/statutes/32/title32ch114sec0.html

No other information is available at time of printing

MARYLAND   MD

Maryland Real Estate Commission: http://www.dllr.state.md.us/license/occprof/recomm.html

<u>*An unlicensed assistant May*</u>

1. Answer the telephone and forward calls to a licensee
2. Submit listings and changes to a multiple listing service
3. Follow up on loan commitments after a contract has been negotiated
4. Assemble documents for closing
5. Secure documents for closing
6. Secure documents (public information) from courthouse, public utilities, etc.
7. Have keys made for company listings
8. Write and place ads for approval of licensee and supervising broker
9. Type contract forms at the direction of and for approval by licensee and supervising broker
10. Computer commission checks
11. Place signs on property
12. Arrange date and time for well/septic inspection, mortgage application, pre-settlement walk-through, and settlement
13. Prepare flyers and promotional information for approval by licensee and supervising broker

14. Act as courier service to deliver documents, pick up keys, etc.
15. Schedule and open house
16. Schedule appointments for licensee to show listed property
17. Accompany a licensee to an open house or showing for security purposes or to hand out preprinted materials

An unlicensed assistant May Not

1. Prepare promotional material or ads without the review and approval of the licensee and supervising broker
2. Show property
3. Answer any questions on listings, title, financing, closing, etc.
4. Discuss or explain a contract, listing, lease, agreement or other real estate document with anyone outside the brokerage
5. Be paid on the basis of real estate activity such as a percentage of commission, or any amount based on listings, sales, etc.
6. Negotiate or agree to any commission split, management fee, or referral fee on behalf of licensee
7. Discuss the attributes or amenities of a property, under any circumstances, with a prospective purchaser or lessee
8. Discuss with the owner of real property the terms and conditions of the real property offered for sale or lease
9. Collect, receive or hold deposit monies, rent, other monies or anything of value received from the owner of the real property or from a prospective purchaser or lessee
10. Provide owners of real property or prospective purchasers or lessees with any advise, recommendations or suggestions as to the sale, leasing of real property to be listed or presently available for sale or lease
11. Hold himself or herself out in any manner, orally or in writing, as being licensed or affiliated with a particular company or real estate broker as a licensee
12. Contact the public concerning the availability of real estate brokerage services unless an inquiry about a specific property is immediately referred to a licensee

MASSACHUSETTS   MA

Massachusetts Board of Registration—Board of Real Estate Brokers and Salesmen:
http://www.mass.gov/dpl/boards/re

No other information is available at time of printing

MICHIGAN   MI

Michigan Real Estate Commission:

http://www.michigan.gov/cis/0,1607,%207-154-10557_12992_14257—-,00.html

An unlicensed assistant May

1. Accompany licensees during the holding of an open house and perform the following functions as "host" or "hostess":
   a. Open the door and greet prospects as they arrive at the open house;
   b. Hand out or distribute prepared printed material;
   c. Have prospects sign a register (guest book) to record names, address and phone numbers for the listing; and
   d. Accompany prospects through the home for security purposes (only the licensee should answer any questions pertaining to the material aspects of the house or its price and terms);
2. Perform strictly clerical tasks; and
3. Function as a courier in picking up or delivering documents on behalf of the employing licensee. [Note: Keys should not be given to unlicensed persons for the purpose of showing a listed property. Brokers are responsible for the properties in their listing inventory and should only give a key to a licensee who is able to show proper I.D. (e.g., valid pocket card or a driver's license with photo).] See, http://michigan.gov/realestatelicense (Frequently Asked Questions—Real Estate).
4. Examples of purely clerical or administrative functions might also include:
   a. setting up or removing signs
   b. scheduling appointments for a licensee
   c. responding to simple questions over the telephone or at open houses by providing basic information that has appeared in a classified newspaper advertisement or in a brochure that has been distributed to the public, such as the location or the price of a home.

An unlicensed assistant MAY NOT:

1. Independently show or demonstrate property to prospective buyers;
2. Make cold calls by telephone or in person to potential listers, purchasers, tenants or landlords;
3. Answer any questions on title insurance, financing or closings;
4. Independently hold open houses for brokers or staff booths at home shows or fairs;
5. Solicit business through telephone prospecting;
6. Give additional information not included in prepared written promotional material that has been distributed to the public (e.g., newspaper ads, flyers, brochures);
7. Represent themselves as an agent for a real estate broker or the owner/seller of property;
8. Have their name printed on business cards or stationery that would imply they are an agent for the real estate broker;

9. Conduct telephone solicitation calls. If John Doe, an unlicensed assistant, calls and indicates he represents ABC Realty, one is led to believe the purpose of the call is to engage in real estate activities. The definition of broker and salesperson in the Code includes on who "lists or Attempts to list." Therefore, a call by an unlicensed assistant identifying him—or herself as a "representative" of a real estate company is an attempt to list even if specific terms are not discussed at that time;

10. Perform any of the acts for which a license is required under Michigan Real Estate License Law. (MCL 339.2501 et seq.)

MINNESOTA   MN

Minnesota Department of Commerce—Real Estate:
http://www.state.mn.us/portal/mn/jsp/content.do?subchannel=-536881740&id=-536881351&agency=Commerce

As stated in the statute below a non-licensed person may disclose information as directed by the broker in written form. Other duties which do not require interaction with the public are not addressed in statute.

Minnesota Statutes 2004,
82.48 Standards of conduct.
Subd. 3. Responsibilities of brokers.

(d) Disclosure of listed property information. A broker may allow any unlicensed person, who is authorized by the broker, to disclose any factual information pertaining to the properties listed with the broker, if the factual information is provided to the unlicensed person in written form by the broker representing or assisting the seller(s).

Should you require additional information please contact the Department at 651-296-2488 or 1-800-657-3602 option number 3 will get you to the real estate division. The above statute may be found at *www.leg.state.mn.us*

MISSISSIPPI   MS

Mississippi Real Estate Commission: http://www.mrec.state.ms.us

Specific information about unlicensed assistants:
http://www.mrec.state.ms.us/notesstory.asp?ID=42

An "Unlicensed Personal Assistant" who works exclusively for a licensee will ordinarily be an employee rather than an independent contractor under Mississippi and Federal tax, unemployment and workers' compensation law. The licensee must follow all applicable laws. The licensee may pay an employee based on a predetermined rate that is agreeable to both parties as long as the assistant's compensation is NOT in any way related to listings or buyers solicited or obtained by the assistant.

The Mississippi Real Estate Commission (MREC) has created a list of activities that cannot be conducted by an unlicensed personal assistant. The list is NOT inclusive and is intended to serve as a guideline.

Unlicensed Assistants May:

1. Provide "general" information about listed properties such as location, availability, and address (without any solicitation on behalf of the assistant).
2. Perform clerical duties, which may include answering the telephone and forwarding calls.
3. Complete and submit listings and changes to a multiple listing service, type contract forms for approval by the licensee and the principal broker, pick-up and deliver paperwork to other brokers and salespersons, obtain status reports on a loan's progress, assemble closing documents and obtain required public information from governmental entities.
4. Write advertising and promotion materials for approval by the licensee and the principal broker, and arrange to place the advertising.
5. Have keys made for listings and place signs on a listed property.
6. Gather information required for a Broker Price Opinion or a Comparative Marketing Analysis.
7. Schedule appointments for the licensee to show a listed property.
8. May be compensated for their work at a predetermined rate that is not contingent upon the occurrence of a real estate transaction. Licensees may NOT share commissions with unlicensed persons who have assisted in transactions by performing any service with respect to a real estate closing.

Unlicensed Assistants May NOT

1. Independently show properties that are for rent or sale.
2. Host an open house, kiosk, home show booth, fair, or hand out materials at such functions UNLESS a licensee is present at all times.
3. Preview, inspect, or determine (measure) the square footage of any property unless accompanied by a licensee.
4. Prepare promotional materials or advertising without the review and approval of a licensee and the principal broker.
5. Negotiate, discuss or explain a contract, listing, lease or any other real estate document with anyone outside the brokerage firm.
6. Answer questions concerning properties listed with the firm, EXCEPT to confirm that a property is listed, to identify the listing broker or sales agent, and to provide such information as would normally appear in a simple, classified newspaper advertisement (location and/or address).
7. Negotiate the amount of rent, security deposit, or other lease provisions in connection with rental property.

8. Open properties for viewing by prospective purchasers, appraisers, home inspectors, or other professionals.

9. Attend pre-closing walk-through or real estate closings unless accompanied by a licensee.

10. Place calls that would require a license such as cold calling, soliciting listings, contacting sellers, buyers or tenants in person or by phone, contacting expired listings, placing marketing calls, or extending open house invitations.

11. Represent themselves as being a licensee or as being engaged in the business of buying, selling, exchanging, renting, leasing, managing, auctioning, or dealing with options on any real estate or the improvement thereon for others.

MISSOURI   MO

Missouri Division of Professional Registration—Real Estate: http://pr.mo.gov/realestate.asp

Specific information about unlicensed assistants:
http://www.pr.mo.gov/boards/realestate/newsletters/1996-03-01.pdf

An unlicensed assistant May

1. Answer the phone and forward calls to a licensee;
2. Submit listings and changes to a multiple listing service;
3. Follow up on loan commitments after a contract has been negotiated;
4. Assemble documents for closings;
5. Secure documents (public information) from the courthouse, sewer district, water district, etc.;
6. Have keys made for company listings;
7. Write ads, flyers, and promotional materials for approval by licensee and supervising broker and place advertising;
8. Record and deposit earnest money, security deposits, and advance rents;
9. Type contract forms for approval by licensee and supervising broker;
10. Monitor licenses and personnel files;
11. Compute commission checks;
12. Place signs on listed property;
13. Order items of routine repair as directed by licensee;
14. Act as a courier to deliver documents, pick up keys, etc.
15. Place routine telephone calls on late rent payments;
16. Schedule appointments for licensee to show listed property.

An unlicensed assistant May Not

1. Be licensed with one firm and do any real estate activities that require a license while working as an assistant in another firm. A licensee who is moonlighting as an assistant in

a second real estate company may do only the activities listed above when working for the second company.

2. Host open houses, kiosks, home show booths or fairs, or hand out materials.

3. Prepare promotional materials or ads without the review and approval of a licensee and the supervising broker;

4. Show property;

5. Answer any questions on listings, title, financing, closing, for anyone outside the firm;

6. Discuss or explain a contract, listing, lease, agreement, or other real estate document with anyone outside the firm;

7. Be paid on the basis of real estate activity, such as a percentage of commission, or an amount based on volume of listings, sales, etc.;

8. Negotiate or agree to a commission, commission split, management fee or referral fee on behalf of a licensee.

MONTANA   MT

Montana Board of Real Estate:
http://www.discoveringmontana.com/dli/bsd/license/bsd_boards/rea_board/board_page.asp

An unlicensed assistant

1. May not access a property in the presence of a potential buyer/tenant unless accompanied by a licensee
   a. This would allow an unlicensed assistant to open property for MLS caravans, appraisers, utility and repair people, home inspectors, deliver flyers, stage a home for showing, take photos, or inspect vacant homes without the presence of a potential buyer/tenant

2. May not contact a prospective seller/buyer/landlord or tenant for the purpose of soliciting a listing agreement; buyers broker agency agreement; property management agreement or rental/lease agreement
   a. This would allow an unlicensed assistant to contact and extend an invitation to an open house.

3. May not attend settlement/closings unless in the presence of a licensee.

4. May not negotiate contracts, rental agreements or leases, buyer/seller agency agreements, listings, title matters, financing or closing issues.
   a. This would allow an unlicensed assistant to provide information which is generally available to the public or which has been disclosed in the marketing process.

5. May not disclose any confidential information.

NEBRASKA   NE

Nebraska Real Estate Commission: http://www.nrec.state.ne.us

An unlicensed assistant May

1. Answer the phone and forward calls to licensees
2. Transmit listings and changes to a multiple listing service
3. Follow up on loan commitments after a contract has been negotiated
4. Assemble documents for closings
5. Secure documents, i.e. public information, from courthouse, sewer district, water district, etc.
6. Have keys made for company listings
7. Write and prepare ads, flyers and promotional information and place such advertising
8. Record and deposit earnest money and other trust funds
9. Type contract forms under direction of licensee
10. Monitor licenses and personnel files
11. Compute commission checks
12. Place and remove signs on property
13. Order items of routine repair as directed by licensee and/or supervising broker
14. Act as courier service to deliver documents, pick up keys, etc.
15. Schedule appointments
16. Measure property if measurements are verified by the licensee
17. Hand out objective written information on a listing, other than at functions such as open houses, kiosks, and home show booths or fairs

An unlicensed assistant May Not

1. Host open houses, kiosks, home show booths or fairs, or hand out materials at such functions
2. Show property
3. Answer any questions on listings, title, financing, closing, etc.
4. Discuss or explain a contract, agreement, listing, or other real estate document with anyone outside the firm
5. Be paid on the basis of real estate activity, such as a percentage of commission, or any amount based on listings, sales, etc.
6. Negotiate or agree to any commission, commission split or referral fee on behalf of a licensee

NEVADA　　NV

Nevada Real Estate Commission: http://www.red.state.nv.us/re_licreq.htm

Specific information about unlicensed assistants: http://red.state.nv.us/bulletin/Bulletin010.pdf

An unlicensed assistant May

The following is a list of *administrative* functions which may be performed by an unlicensed assistant *under the direction of a licensee and supervising broker*:

1. Answer phones, forward calls to or take messages for licensees.
2. Transmit listings and changes to a multiple listing service.
3. Follow up on *administrative* aspects of loan commitments *after* a contract has been negotiated by a licensee.
4. Draft and assemble transaction documents, draft correspondence, do office filing, develop and maintain mailing lists, and perform other clerical duties for a licensee.
5. Research, secure documents, make and deliver copies from public records.
6. Have keys made for company listings.
7. Act as a courier service to deliver documents, pick up keys, etc.
8. Write and prepare newsletters, advertising, flyers, and promotional information and place such advertising after approval by licensee and supervising broker. *Remember: NRS 645.315 requires that ALL advertising must be done under the direct supervision of and in the name of the brokerage.*
9. Perform bookkeeping, record and deposit trust funds under direction of the broker.
10. Monitor licenses and personnel files.
11. Place and/or remove signs on property.
12. Accept rental payments and issue receipts at the broker's place of business.
13. Witness signatures.
14. Schedule routine inspections and arrange for routine repairs on property.

An unlicensed assistant May Not

1. Meet with clients to obtain or renew brokerage agreements or property management agreements.
2. Negotiate or agree to any commission, commission split, management fee or referral fee on behalf of a licensee or receive a referral fee from a licensee.
3. Provide advice or guidance to a client or consumer regarding a real estate contract, brokerage agreement, property management agreement, title, financing, closing or other real estate document.
4. Show property or provide clients or consumers information on listings.
5. Answer any questions about a listing, including asking price, square footage, age of structure.
6. Give listing presentations, interview buyers or present or negotiate offers.
7. Contact or solicit prospective sellers or buyers, landlords or tenants, including scheduling appointments as a result of a telemarketing survey asking any of those parties if they would like to speak with a licensee about their real estate questions.

May an unlicensed assistant host an open house?

Yes, but care must be taken that the unlicensed assistant does not *show* the property to prospective purchasers. That means an unlicensed assistant may welcome visitors, hand-out brochures prepared by the licensee and serve refreshments at an open house, but all inquiries about the listing must be referred to a licensee. The host must NOT point out features of the home or neighborhood to visitors, but may distribute flyers or brochures prepared by a licensee that describe the property. The same rules for what an unlicensed assistant can and cannot do apply to hosting an open house.

NEW HAMPSHIRE NH

New Hampshire Real Estate Commission: http://www.nh.gov/nhrec

An unlicensed assistant May:

1. Give general information about listed properties such as location, availability and price (again, without any solicitation on behalf of the assistant);
2. Perform clerical duties, which may include answering the telephone and forwarding calls;
3. Fill out and submit listings and changes to multiple listing services;
4. Type contract forms for approval by licensee and supervising broker;
5. Pick up and deliver paperwork to other brokers and salespersons;
6. Follow-up on loan commitments after a contract has been negotiated, and pick up and deliver loan documents requiring signatures;
7. Obtain status reports on a loans progress;
8. Assemble closing documents;
9. Obtain required public information from the Registry of Deeds, public utilitiles,etc.;
10. Write advertising for approval by the licensee and supervising broker, and arrange to place the advertising;
11. Have keys made for company listings, and place signs on listed property;
12. Attend open houses to provide security, and hand out pre-approved promotional material;
13. Gather information required for a Comparative Market Analysis;
14. Schedule appointments for licensee to show a listed property;
15. Chauffeur clients to view properties which will be shown by the licensee.
16. A good general rule is that an unlicensed assistant should have very little contact with the buyer or seller beyond providing secretarial assistance or factual information on listings.

NEW JERSEY   NJ

New Jersey Department of Banking and Insurance—Real Estate:
http://www.state.nj.us/dobi/remnu.shtml

An unlicensed assistant May

1. Answer phones and forward calls.
2. Process and submit listings and changes to a MLS system.
3. Follow-up on loan applications after contracts have been fully executed.
4. Set up file procedures, track and secure documents, etc.
5. Have keys made for company listings at the direction of a licensee.
6. Write ads for approval of a licensee, place ads as directed.
7. Keep records of, and deposit payments of earnest money, security deposits and rent.
8. Type contract forms for approval of a licensee.
9. Monitor files and report findings to a licensee.
10. Compute commission checks.
11. Place signs on properties.
12. Order items or inspections as directed by a licensee.
13. Prepare flyers and promotional material for approval by licensee.
14. Act as a courier for delivering documents or picking up keys etc. (Licensee is responsible for delivery of contracts or closing materials).
15. Schedule appointments with the seller or seller's agent in order for a licensee to show listed property.

An unlicensed assistant May Not

1. Make cold calls by telephone or in person to potential listers, purchasers, tenants, or landlords.
2. In the absence of a licensee, host open houses, booths at home shows, malls or fairs, or distribute promotional literature at such locations.
3. Prepare promotional material or ads without the review and approval of a licensee.
4. Show property.
5. Answer any questions on listings, title, financing or closings from either the public or other licensees.
6. Discuss or explain a contract, listing, lease agreement or other real estate document with anyone outside the firm.
7. Work as a licensee/secretary in one firm and do real estate related activities with that firm, while licensed with another firm.
8. Negotiate or agree to any commission, commission split, management fee or referral fee on behalf of a licensee.
9. In addition, the compensation of a personal assistant or secretary should not be based on the success of their activity, i.e. a percentage of commission, but should be directly related to the duties the non-licensee is performing. If a licensee is using another licensee to act as their personal assistant/secretary, both should be aware that they are employees or independent contractors of their broker and compensation must be paid by the broker.

All licensees are cautioned to research and adhere to Federal and State Income Tax and Employment requirements.

In the course of revisiting this issue the Commission considered additional tasks performed by unlicensed persons that were not previously addressed. These included the placing of routine calls on late rent payments and being present at inspections for security reasons.

Pursuant to N.J.S.A. 45:15-3, the actions which require licensure as a real estate broker include "collecting, or offering or attempting to collect rent for the use of real estate." On the basis of this statuary provision the Commission concluded that only licensed individuals may make telephone calls for purpose of collecting or attempting to collect late rent payments.

With regard to inspections, because it is highly likely that during the course of an inspection questions will be raised by the prospective purchaser and/or the owner of the property which only a licensee would be qualified to answer, it was determined that an unlicensed individual should not be present during such inspections in the absence of a licensee. There would clearly be no impropriety where, for security reasons, a licensed individual requested their unlicensed assistant to accompany them to an inspection.

A final issue reviewed involved to what extent, if at all, unlicensed persons present at open houses may respond to questions that may be answered with objective responses gleaned from form preprinted objective information, for example, how many bedrooms or bathrooms a house has. The 1992 list of functions which unlicensed individuals may not perform contained the following item: "Answer any questions on listings, title, financing or closings from either the public or licensees." (emphasis added). It remains the position of the Commission that unlicensed individuals should not answer any questions on listings, even if the questions only inquire about objective information which is contained in pre-printed material about the property. It is helpful to recall that unlicensed persons may not host open houses in the absence of a licensee. Therefore, the unlicensed individual can refer the person making the inquiry to the written material wherein the answer is contained or to the licensee in attendance.

Adhering to these guidelines will enable licensees and unlicensed individuals to avoid potential violations.

NEW MEXICO   NM

New Mexico Regulation and Licensing Division—Real Estate: http://rld.state.nm.us/b&c/recom

Specific information about unlicensed assistants:
http://www.nmcpr.state.nm.us/nmac/parts/title16/16.061.0021.htm

An unlicensed assistant May

1. Obtain information pursuant to written instructions from the Responsible Person from public records, a Multiple Listing Service, Listing Exchange or from third party sources including, but not limited to, surveyors, banks, appraisers and title companies.
2. Host and/or distribute literature at an open house under the following conditions:
   a. An Unlicensed Assistant does not discuss, negotiate or solicit offers for the property or provide any information other than printed material prepared and approved by the Responsible Person; and
   b. The Responsible Person is present at the open house where the Unlicensed Assistant is located;
   c. All inquiries are referred to the Responsible Person or other Licensee.
3. Disseminating and distributing information prepared and approved by the Responsible Person.
4. Picking up and delivering paperwork to Licensees other than the Responsible Person
5. Picking up and delivering paperwork to sellers or purchasers after a contract has been executed if the paperwork has already been reviewed and approved by the Responsible Person, without answering any questions or providing any opinions or advice to the recipient of the paperwork. All substantive questions must be referred to the Responsible Person
6. Writing advertisements, flyers, brochures, and other promotional materials for the approval of the Responsible Person, and placing classified advertisements approved by the Responsible Person.
7. Placing or removing signs on real property as directed by the Responsible Person.
8. Ordering repairs as directed by the Responsible person.
9. Receiving and depositing funds, maintaining books and records, while under the supervision of the Responsible Person.
10. Typing or word processing documents, including purchase and listing agreements, prepared by the Responsible Person.

An unlicensed assistant May Not

1. Preparing legal documents such as listing and sales contracts.
2. Interpreting documents, offering opinions or advice.
3. Disseminating and distributing information, unless the information is in writing and is prepared and approved by the Responsible Person.
4. Obtaining personal or property information from a Client or Customer of the Responsible Person except when acting as a coordinator directed by the Responsible Person by gathering and following up on information and the status of matters pertaining to the Transaction after a contract has been executed.
5. Picking up from or delivering to customers or clients financial documents prepared by title companies, lenders or other third persons for the purpose of obtaining signatures.

6. Attending a closing without the Responsible Person present.
7. Representing himself or herself as being a Licensee or as being engaged in the business of buying, selling, exchanging, renting, leasing, managing, auctioning or dealing with options on any real estate or the improvements thereon for others.
8. Telephone solicitation of any kind designed to procure transactions requiring licensure under Section 61-29-1 et. Seq. NMSA 1978, including, but not limited to, procuring buyers, sellers, listings or appointments for listing presentation.

NEW YORK   NY

New York Real Estate Commission: http://www.dos.state.ny.us/lcns/realest.html

Real Estate Board of New York: http://www.rebny.com

An unlicensed assistant May

1. Answer the phone and forward calls to licensees
2. Transmit listings and changes to a multiple listing service
3. Follow up on loan commitments after a contract has been negotiated
4. Assemble documents for closings
5. Secure documents, i.e. public information, from courthouse, sewer district, water district, etc.
6. Have keys made for company listings
7. Write and prepare ads, flyers and promotional information and place such advertising
8. Record and deposit earnest money and other trust funds
9. Type contract forms under direction of licensee
10. Monitor licenses and personnel files
11. Compute commission checks
12. Place and remove signs on property
13. Order items of routine repair as directed by licensee and/or supervising broker
14. Act as courier service to deliver documents, pick up keys, etc.
15. Schedule appointments
16. Measure property if measurements are verified by the licensee
17. Hand out objective written information on a listing, other than at functions such as open houses, kiosks, and home show booths or fairs

An unlicensed assistant May Not

1. Host open houses, kiosks, home show booths or fairs, or hand out materials at such functions
2. Show property
3. Answer any questions on listings, title, financing, closing, etc.
4. Discuss or explain a contract, agreement, listing, or other real estate document with anyone outside the firm

5.   Be paid on the basis of real estate activity, such as a percentage of commission, or any amount based on listings, sales, etc.

6.   Negotiate or agree to any commission, commission split or referral fee on behalf of a licensee

NORTH CAROLINA   NC

North Carolina Real Estate Commission: http://www.ncrec.state.nc.us

No other information available at time of printing

NORTH DAKOTA   ND

North Dakota Real Estate Commission:
http://www.governor.state.nd.us/boards/boards-query.asp?Board_ID=93

Licensees, both brokers and salespersons, often use unlicensed persons, either employed or contracted, to perform various tasks related to a real estate transaction which do not require a license. Such persons, for example, are used as personal assistants, clerical support staff, closing secretaries, etc.

The North Dakota Real Estate License Law prohibits unlicensed persons from negotiating, listing, or selling real property. It is, therefore, important for employing brokers and other licensees using such persons to carefully restrict the activities of such persons so that allegations of wrongdoing under the law or rules can be avoided.

Licensees should not share commission with unlicensed persons acting as assistants, clerical staff, closing secretaries, etc. The temptation for such unlicensed persons, in such situations, to go beyond what they can do and negotiate or take part in other prohibited activities is greatly increased when their compensation is based on the successful completion of the sale.

In order to provide guidance to licensees regarding which activities relating to a real estate transaction unlicensed persons can and cannot perform, the Commission has established the following guidelines:

Activities which CAN be performed by unlicensed persons who, for example, act as personal assistants, clerical support staff, closing secretaries, etc., include, but are not necessarily limited to:

1.   Answer the phone and forward calls to licensees.
2.   Transmit listing data and changes to a multiple listing service.
3.   Follow up on loan commitments after a contract has been negotiated.
4.   Assemble documents for closings.
5.   Secure documents, i.e. public information, from courthouse, register of deeds, or tax office.
6.   Have keys made for the firm's listings.
7.   Record and deposit earnest money, security deposits, and other trust monies.

8.   Type offers, contracts and leases under the direction of the licensee.

9.   Check license renewal and personnel files for the brokers and salespersons with the firm.

10. Compute commission checks and act as bookkeeper for the firm's operating bank accounts.

11. Place and remove signs on property at the direction of a broker or salesperson with the firm.

12. Order and supervise routine and minor repairs as directed by the licensee and/or supervising broker.

13. Act as a courier to deliver or pick up documents, keys, etc.

14. Schedule appointments.

15. Measure property, if licensee verifies measurements.

16. Write and prepare ads, flyers and promotional information and place such advertising.

17. Hand out objective written information on a listing, other than at functions such as open houses, kiosks, and home show booths or fairs.

Activities which CANNOT be performed by unlicensed persons who, for example, act as personal assistants, clerical support staff, closing secretaries, etc., include, but are not necessarily limited to:

1.   Make solicitations by telephone or in person to potential listers and purchasers.

2.   Show property.

3.   Host open houses, kiosks, home show booths, or fairs or hand out materials at such functions.

4.   Answer any questions on listings, title, financing, closing, etc.

5.   Discuss or explain listings, offers, contracts, or other similar matters with persons outside the firm.

6.   Be paid on the basis of real estate activity, such as a percentage of commission, or any amount based on listings, sales, etc.

7.   Negotiate or agree to any commission, commission split or referral fee on behalf of a licensee.

8.   Act as a "go-between" with a seller and buyer such as when an offer is being negotiated.

Employing brokers, whether they are employing unlicensed persons or whether licensees under their supervision are using unlicensed persons as personal assistants or the like, are responsible for assuring that such unlicensed persons are not involved in activities which require a license and/or activities as stated in these guidelines. Brokers should establish guidelines for the use of unlicensed persons and procedures for monitoring their activities. It is the responsibility of the employing broker to assure that unlicensed persons, either employed or contracted by licensees under his/her supervision, are not acting improperly.

OHIO   OH

Ohio Real Estate Commission: http://www.com.state.oh.us/real/remain.htm

An unlicensed assistant May

1. Perform duties with care secretarial in nature
2. Call to schedule appointments
3. Call the owner of a property listed by the brokerage to schedule a showing, closing or inspection
    a. The dialogue of the conversations should be limited to setting an appointment and may NOT focus on making representations about the services offered by the firm.
4. Clerical support
5. Filing
6. Taking messages

An unlicensed assistant May Not

1. Provide information over the phone to prospective buyers and/or real estate agents.
2. Answer questions concerning a property, such as asking price, address or number of bedrooms.
3. Direct or attempt to procure a prospect
4. Cold calling to determine the party's interest in listing or re-listing a property.
5. Gather information on an owner's house or home they may be looking for.
6. Provide property-related information.
7. Staff an open house

OKLAHOMA   OK

Oklahoma Real Estate Commission: http://www.state.ok.us/~orec

An unlicensed assistant May not:

1. Prepare Comparative Market Analysis.
2. Communicate with Attorneys.
3. Measure Houses (Certified or Licensed Individual).
4. Give Listing Presentations.
5. Communicate with Buyers and Sellers.
6. Make Cold Calls.
7. Check Listings to Verify Information.
8. Interview Buyers.
9. Answer Questions on Listings, or Any Item Pertaining to Transaction.
10. Submit Listings.
11. Give Out Information on Listing Properties.
12. Paid a Percentage or Share of a Commission.
13. Show Property.
14. Hang Door Hangers in Neighborhoods.
15. Relocate Families to Temporary Housing.

16. Do Personal Prospecting.
17. Communicate with For Sale By Owner.
18. Communicate with Expired Listings.
19. Coordinate Buyer Seminar.
20. Host Open Houses, Kiosks, Home Show, Booths or Fair, or Hand Out Materials.
21. Schedule Listing Appointments (with telemarketing survey asking home owners if they would like to speak with a licensee about their home).
22. Manage Rental Properties.
23. Communicate with Out-of-Towners.

An unlicensed assistant May:

1. Design Feature Sheets at Direction of Broker.
2. Prepare Net to Seller Sheets at Direction of Broker.
3. Answer the Telephone. Forward Calls. Take Messages.
4. Make Appointments for Licensees on Transaction Commitments or Incoming Calls From Prospects.
5. Arrange Loan Applications.
6. Write Newsletters.
7. Write, Design and Place Ads for Approval by Supervising Broker.
8. Develop and Maintain Mailing Lists—Clerical Duties Only.
9. Process Loans.
10. Coordinate "Move-In Day" after Closing.
11. Schedule Inspections.
12. Communicate with Cooperative Licensees to Set Up Showing Appointments.
13. Run the Computer to Obtain Information for Licensee.
14. Input Information into Computer.
15. Shop Financing by Checking Rates with Lenders on Specific Buyers and Communicate to Licensee.
16. Order, Pick Up and Deliver Documents Pertaining to a Transaction at Direction of Licensee Without Contacting the Buyer or Seller.
17. Identify Expired Listings on Computer.
18. Communicate with Service Companies at the Request of Licensee.
19. Provide Personal Accounting for Licensee (personal bookkeeping).
20. Inspect Listing Forms and Re-supply Handouts on Yard Signs.
21. Pay Personal Bills of Licensee.
22. Maintain Past Buyer and Seller List (input data or check data on computer for licensee's future solicitation: but not by personally contacting client).
23. Scan and Clip News.
24. Follow Up on Loan Commitments at Direction of Licensee.
25. Record and Deposit Money to Be Placed in Escrow at Direction of Sponsoring Broker.

26. Type Contract Forms at Direction of Licensee.
27. Monitor License and Personnel Files.
28. Issue Computer Commission Checks at the Direction of the Sponsoring Broker.
29. Order Items or Routine Repair as Directed by Licensee.
30. Act as a "Security Host" Only at an Open House—All Questions Must be Referred to Licensee. (In the best interest of the public, the Commission suggests that a licensee be available to answer questions).
31. Put Up and Pick Up Signs.
32. Have Keys Made for Company Listings. Pick Up Keys at Direction of Broker.
33. Take Pictures of Houses at Direction of Licensee.
34. Assemble Documents for Closing at Direction of Licensee.
35. Direct Mailing.
36. Send Ads to Seller at Direction of Licensee.
37. Accept Rent and Provide Receipt.
38. Type Rental Agreements for the Licensee to Have Signed.

OREGON   OR

State of Oregon Real Estate Agency: http://www.rea.state.or.us

Specific information about unlicensed assistants: http://www.leg.state.or.us/ors/696.html

An unlicensed assistant May Not

"Professional real estate activity" means any of the following actions, when engaged in for another and for compensation or with the intention or in the expectation or upon the promise of receiving or collecting compensation, by any person who:

1. Sells, exchanges, purchases, rents or leases real estate.
2. Offers to sell, exchange, purchase, rent or lease real estate.
3. Negotiates, offers, attempts or agrees to negotiate the sale, exchange, purchase, rental or leasing of real estate.
4. Lists, offers, attempts or agrees to list real estate for sale.
5. Offers, attempts or agrees to perform or provide a competitive market analysis or letter opinion, to represent a taxpayer under ORS 305.230 or 309.100 or to give an opinion in any administrative or judicial proceeding regarding the value of real estate for taxation. Such activity performed by a state certified appraiser or state licensed appraiser is not professional real estate activity.
6. Auctions, offers, attempts or agrees to auction real estate.
7. Buys, sells, offers to buy or sell or otherwise deals in options on real estate.
8. Engages in management of rental real estate.

9. Purports to be engaged in the business of buying, selling, exchanging, renting or leasing real estate.
10. Assists or directs in the procuring of prospects, calculated to result in the sale, exchange, leasing or rental of real estate.
11. Assists or directs in the negotiation or closing of any transaction calculated or intended to result in the sale, exchange, leasing or rental of real estate.
12. Except as otherwise provided in ORS 696.030 (1)(L), advises, counsels, consults or analyzes in connection with real estate values, sales or dispositions, including dispositions through eminent domain procedures.
13. Advises, counsels, consults or analyzes in connection with the acquisition or sale of real estate by an entity if the purpose of the entity is investment in real estate.
14. Performs real estate marketing activity as described in ORS 696.600.

PENNSYLVANIA   PA

Pennsylvania Real Estate Commission:
http://www.dos.state.pa.us/bpoa/cwp/view.asp?a=1104&Q=433107

An unlicensed assistant or individual may not perform any duties that require licensure.

An unlicensed assistant May Not:

1. Host an open house to be attended by the public.
2. Explain or interpret information on the listing sheets to licensees or the public, whether in person, over the telephone or over the Internet.
3. Provide information on the style of the home, location of the home, lifestyle or amenities available.
4. Tour/show a model to the public.
5. Discuss prices, price ranges or mortgage rates with the public.
6. Engage in telemarketing, excluding cemetery sales.

An unlicensed assistant May for example:

1. Furnish information from listing sheets to other real estate licensees/offices.
2. Host an open house to be attended by only real estate licensees.
3. Prepare and distribute brochures about listed properties to other real estate offices.
4. Erect "for sale" signs.
5. Maintain lock boxes.
6. Perform general clerical duties.

Unlicensed assistants or individuals may be employed, supervised and directly compensated by a broker, associate broker or salesperson. Licensees who permit unlicensed assistants or individuals to perform services that require licensure (i.e. those listed in the first category), may be disciplined for aiding and abetting unlicensed practice.

RHODE ISLAND   RI

Rhode Island Real Estate Commission: http://www.dbr.state.ri.us/real_estate.html

No other information available at time of printing

SOUTH CAROLINA   SC

South Carolina Real Estate Commission: http://www.llr.state.sc.us/POL/RealEstateCommission/

Specific information on unlicensed assistants:
http://www.llr.state.sc.us/POL/RealEstateCommission/Black%20Book—2004.pdf page 17

An unlicensed assistant May

1. Perform maintenance;
2. Perform clerical or administrative support;
3. Perform collection of rents which are made payable to the owner or real estate company;
4. Show rental units to prospective tenants;
5. Furnish published information;
6. Provide applications and lease forms;
7. Receive applications and leases for submission to the owner or the licensee for approval.

An unlicensed assistant May Not

1. Discuss, negotiate, or explain a contract, listing, buyer agency, lease, agreement, or other real estate document;
2. Vary or deviate from the rental price or other terms and conditions previously established by the owner or licensee when supplying relevant information concerning the rental of property;
3. Approve applications or leases or settle or arrange the terms and conditions of a lease;
4. Indicate to the public that the unlicensed individual is in a position of authority which has the managerial responsibility of the rental property;
5. Conduct or host an open house or manage an on-site sales office;
6. Show real property;
7. Answer questions regarding company listings, title, financing, and closing issues, except for information that is otherwise publicly available;
8. Discuss, negotiate, or explain a contract, listing, buyer agency, lease, agreement, or other real estate document;

9. Be paid solely on the basis of real estate activity including, but not limited to, a percentage of commission or any amount based on the listing or sales compensation or commission;

10. Negotiate or agree to compensation or commission including, but not limited to, commission splits, management fees, or referral fees on behalf of a licensee;

11. Engage in an activity requiring a real estate license as required and defined by this chapter.

SOUTH DAKOTA   SD

South Dakota Real Estate Commission: http://www.state.sd.us/sdrec

An unlicensed assistant May:

1. Deliver documents and pick up keys.
2. Answer the telephone and forward calls.
3. Secure public information from courthouses, utility districts, etc
4. Provide courier services.
5. Schedule appointments with other offices, existing clients, or customers.
6. Place signs on property.
7. Type forms for approval by licensee and supervising broker.
8. Write ads for approval of licensee and supervising broker, and place classified advertising.
9. Assemble documents for closing.
10. Hand out objective, written information on a listing.

An unlicensed assistant May Not

1. Show property to prospective buyers.
2. Solicit by telephone or in person potential sellers, purchasers, ten-ants or landlords.
3. Answer questions on title insurance, financing or closings.
4. Host open houses for licensees or the public, or staff booths at home shows or fairs.
5. Give additional information not included in prepared written promotional material that has been distributed to the public.
6. Represent himself or herself as an agent for a real estate broker or the owner/seller of a property.
7. Negotiate or discuss the terms of a sale.
8. Be paid on the basis of real estate activity, such as a percentage of commission, or any amount based on listings, sales, etc.
9. Act as a go-between with a seller and buyer.
10. Answer questions concerning properties listed with the firm, except to confirm that the property is listed and identify the listing broker or salesperson.
11. Solicit bidders for real estate sold at auction.

TENNESSEE TN

Tennessee Real Estate Commission: http://www.state.tn.us/commerce/boards/trec

Specific information on unlicensed assistants:
http://www.state.tn.us/commerce/boards/trec/pdf/FAQRevised092204.pdf

An unlicensed assistant May

1. Answer the phone, forward calls and give information contained only on the listing agreement as limited by the broker
2. Fill out and submit listings and changes to any multiple listing service
3. Follow up on loan commitments after a contract has been negotiated and generally secure status reports on the loan progress
4. Assemble documents for closing
5. Secure public information from courthouses, utility districts, etc.
6. Have keys made for company listings
7. Write ads for approval of licensee and principal broker and place classified advertising
8. Receive, record and deposit earnest money, security deposits and advance rents under the direct supervision of the principal broker
9. Type contract forms for approval by licensee and principal broker
10. Monitor licenses and personnel files
11. Compute commission checks
12. Place signs on property
13. Order repairs as directed by licensee
14. Prepare fliers and promotional information for approval by licensee and principal broker
15. Deliver documents and pick up keys
16. Place routing telephone calls on late rent payments
17. Schedule appointments for licensee to show listed property
18. Gather information for a comparative market analysis
19. Hand out objective, written information on a listing or rental
20. Give a key to a prospect or unlock a property
21. Disclose the current sales status of a listed property

An unlicensed assistant May Not

1. Make cold calls by telephone or in person to potential listers or purchasers
2. Show properties for sale and/or lease to prospective purchasers
3. Host public open houses, licensee open houses, home show booths or fairs
4. Answer questions concerning properties listed with the firm except only that information contained on the listing agreement as limited by the principal broker

5. Prepare promotional material or advertising of properties for sale or lease without the approval of the principal broker

6. Discuss or explain listings, offers, contracts, or other similar matters with persons outside the firm

7. Be paid on the basis of real estate activity such as percentage of commission or any amount based on listings, sales, etc.

8. Act as a go-between with a seller and buyer such as when an offer is being negotiated

9. Negotiate or agree to any commission split or referral fee on behalf of a licensee

TEXAS   TX

Texas Real Estate Commission: http://www.trec.state.tx.us

An unlicensed assistant May/May Not

1. Confirm information concerning the size, price and terms of property advertised.

   a. Taken together, this means that an unlicensed person may, after identifying himself or herself as an unlicensed person, confirm information previously advertised to callers or persons dropping by.

   b. The unlicensed person should not give information about properties other than that inquired about, and should refer any requests for information regarding other properties to a licensed agent. For example, the assistant might confirm that a particular property called about has three bedrooms and one bath, as previously advertised; however, the assistant may not attempt to identify properties which instead have two baths and bring these to the attention of the caller. Such questions must be referred to a licensee.

   c. The assistant should not attempt to "qualify" the caller in any respect.

2. The assistant should not attempt to "qualify" the caller in any respect.

3. Many other duties that are administrative in nature can be safely performed, such as inputting data into a computer or typing contracts, but, only as specifically directed by a licensee.

4. Support personnel can order supplies, schedule maintenance, and all the other things that are involved in keeping the office open.

5. Bookkeeping and office management functions may be performed by an unlicensed assistant,

6. Training or motivating personnel,

7. Those tasks dealing with office administration and personnel matters.

Commission Rule 535.2(c) notes that who a broker designates to sign checks in the brokerage is not regulated by the Commission. Thus, an unlicensed person may serve as bookkeeper for the company and handle personnel matters. Such an office manager may also serve as a trainer. However,

Commission Rule 535.1(g) further states that an unlicensed person may not direct or supervise agents in their work as licensees.

Therefore, an unlicensed person may not direct or advise agents in their attempts to help others buy, sell, or lease property. They may not review contracts, or help make "deals" work. These tasks are properly conducted only by licensed persons.

UTAH   UT

Utah Real Estate Commission: http://www.commerce.utah.gov/dre/relicensing.html

Specific information on unlicensed assistants:
http://www.commerce.utah.gov/dre/Rules/Real_Estate/r162-006.rtf

6.2.14. Personal Assistants. With the permission of the principal broker with whom the licensee is affiliated, the licensee may employ an unlicensed individual to provide services in connection with real estate transactions which do not require a real estate license, including the following examples: *4/23/98*

> (a) Clerical duties, including making appointments for prospects to meet with real estate licensees, but only if the contact has been initiated by the prospect and not by the unlicensed person; *6/24/92*

> (b) At an open house, distributing preprinted literature written by a licensee, so long as a licensee is present and the unlicensed person furnishes no additional information concerning the property or financing and does not become involved in negotiating, offering, selling or filling in contracts; *4/23/98*

> (c) Acting only as a courier service in delivering documents, picking up keys, or similar services, so long as the courier does not engage in any discussion of, or filling in of, the documents; *3/3/94*

> (d) Placing brokerage signs on listed properties; *6/24/92*

> (e) Having keys made for listed properties; and   *6/24/92*

> (f) Securing public records from the County Recorders' Offices, zoning offices, sewer districts, water districts, or similar entities. *3/3/94*

6.2.14.1. If personal assistants are compensated for their work, they shall be compensated at a predetermined rate which is not contingent upon the occurrence of real estate transactions. Licensees may not share commissions with unlicensed persons who have assisted in transactions by performing the services listed in this rule. *9/14/92*

6.2.14.2. The licensee who hires the unlicensed person will be responsible for supervising the unlicensed person's activities, and shall ensure that the unlicensed person does not perform activity which requires a real estate license. *9/14/92*

6.2.14.3. Unlicensed individuals may not engage in telephone solicitation or other activity calculated to result in securing prospects for real estate transactions, except as provided in R162-6.2.14.(a) above. *9/14/92*

VERMONT   VT

Vermont Real Estate Commission: http://vtprofessionals.org/opr1/real_estate

No other information available at time of printing

VIRGINIA   VA

Virginia Real Estate Commission: http://www.state.va.us/dpor/apr_main.htm

An unlicensed assistant May

1.   Answer the phone and forward calls to a licensee;
2.   Submit listings and changes to a MLS;
3.   Follow up on loan commitments after a contract has been negotiated;
4.   Assemble documents for closings;
5.   Secure documents from courthouses, public utilities offices, etc.;
6.   Have keys made for company listings;
7.   Write ads for approval of licensee and supervising broker and place advertising;
8.   Record and deposit earnest money, security deposits and advance rents;
9.   Type contract forms for approval by licensee and supervising broker;
10.   Monitor licenses and personnel files;
11.   Compute commission checks;
12.   Place signs on properties;
13.   Order items of routine repair;
14.   Prepare flyers and promotional information for approval by licensee and supervising broker;
15.   Act as a courier service to deliver documents, pick up keys
16.   Place routine telephone calls on late rental payments;
17.   Schedule appointments for showings;

An unlicensed assistant MAY NOT:

1.   Prepare promotional materials or ads without the review and approval of the licensee and supervising broker
2.   Show property;

3. Answer any question on listings, title, financing, closings, etc.

4. Discuss or explain a contract, listing, lease, agreement or other real estate document with anyone outside of the firm;

5. Work as a licensee/secretary in one firm and do real estate related activities within another form;

6. Be paid on the basis of real estate activity, such as a percentage of commissions, or any amount based on listings, profits, etc.;

7. Negotiate or agree to any commission, commission split, management fee or referral fee on behalf of a licensee.

WASHINGTON   WA

Washington Real Estate Commission: http://www.dol.wa.gov/realestate/refront.htm

Specific information on unlicensed assistants: http://www.dol.wa.gov/realestate/reunlic.htm

Unlicensed assistants who work under the direct instructions and supervision of a licensed broker and/or salesperson may perform the following tasks and duties:

1. May be a greeter at open houses/model units and distribute pre-printed promotional literature and provide security.

2. May act as a courier in delivering documents, picking up keys, or similar services.

3. Perform clerical duties such as typing, answering the telephone, forwarding calls, and scheduling appointments for licensees.

4. Submit forms and changes to multiple listing services. Obtain status reports on loan progress and credit reports, etc.

5. Follow-up on loan commitments after a contract has been negotiated; and pick up and deliver loan documents.

6. Obtain public information from sources like government offices, utility companies, title companies, etc.

7. Write and place advertising.

8. Make keys, install lock boxes, and place/remove signs on property.

9. Gather information for comparative market analysis.

10. Transport people to properties and/or around areas of interest but may not show, answer questions, or interpret information regarding property, price or condition.

11. Perform accounting and collection functions such as collecting rents, recording and depositing earnest moneys, security deposits, rental funds and/or computing commission checks.

12. Order or perform items of repair and/or maintenance.

13. Provide information pertaining to the characteristics of real estate or a business opportunity and the terms or the conditions of a transaction only if that information is prepared in writing and approved in advance by a licensee.

Unlicensed assistants May not:

1. Engage in any conduct which is "used, designed or structured" to procure prospects.
2. Show properties, answer questions, or interpret information regarding property, price, or condition.
3. Interpret information regarding listings, titles, financing, contracts, closings or other information relating to a transaction.
4. Conduct telemarketing or telephone canvassing to schedule appointments in order to seek clients.
5. Fill in legal forms or negotiate price and/or terms.
6. Perform any act with the intent to circumvent or which results in the circumvention of the real estate licensing law, RCW 18.85 or the administrative rules in 308-124 WAC

WEST VIRGINIA   WV

West Virginia Real Estate Commission: http://www.wvrec.org

Specific information on unlicensed assistants: http://www.wvrec.org/Law05.pdf see WV Code §30-40-5(c).

No other information is available at time of printing

WISCONSIN   WI

Wisconsin Real Estate Commission: http://drl.wi.gov/prof/rebr/def.htm

No other information is available at time of printing.

WYOMING   WY

Wyoming Real Estate Commission: http://realestate.state.wy.us

An unlicensed assistant may

1. Perform clerical duties, office filing.
2. Draft a document for approval by the licensee.
3. Place or remove signs.
4. Witness signatures.
5. Perform company bookkeeping.
6. Arrange for repairs on rental property.
7. Draft correspondence for approval by licensee.
8. Make and deliver copies of any public records.
9. Answer the telephone, forward calls, take messages, make appointments.
10. Write newsletters.

11. Write, design and place ads for approval by the responsible broker.
12. Develop and maintain mailing lists—clerical duties only.
13. Schedule inspections.
14. Gather information for listing.
15. Accept rental payments and issue receipts.
16. Provide access to a property and hand out preprinted, objective information, so long as no negotiating, offering, selling or contracting is involved.
17. Distribute preprinted, objective information at an open house, so long as no negotiating, offering, selling or contracting is involved.
18. Distribute information on listed properties.
19. Deliver paperwork to other brokers.
20. Deliver paperwork to sellers or purchasers, if such paperwork has already been reviewed by a broker.
21. Deliver paperwork requiring signatures in regard to financing documents that are prepared by lending institutions.
22. Prepare market analyses for sellers or buyers on behalf of a broker, but disclosure of the name of the preparer must be given, and it must be submitted by the broker.

An unlicensed assistant MAY NOT:

1. Host an open house.
2. Provide advice or guidance to a consumer with regards to a listing contract, property management contract, contract to purchase, etc.
3. Issue receipts for earnest money or sign for receipt of sales contracts.
4. Meet with owners to obtain or renew listing agreements, property management agreements, etc.
5. Present or negotiate offers.
6. Enter into a rental contract on behalf of the licensee.
7. Communicate with consumers about real estate transactions.
8. Collect rents or agree to collect rent for the use of real estate.
9. Receive a referral fee from licensee.
10. Be paid a percentage of the commission received by licensee.
11. Assist or direct in the procuring or prospects calculated to result in the sale, exchange, lease or rental of real estate. (Includes making "cold calls.")
12. Give listing presentations.
13. Interview buyers.
14. Answer questions on listings or any item pertaining to any transaction.
15. Orally provide information on listed property.
16. Show property.
17. Communicate with for sale by owner.

18. Schedule listing appointments as a result of telemarketing survey asking homeowners if they would like to speak with a licensee about their home.

APPENDIX B

RapidListings MLS/IDX Coverage

NORTHERN CALIFORNIA	
Central Valley Association of Realtors (CACENVALL)	CA
CA—Bay Area Real Estate Information Services (BAREIS)	San Francisco North Bay
Marin Association of Realtors	
Napa Association of Realtors	
North Bay Association of Realtors	
Northern Solano County Association of Realtors	
Solano Association of Realtors	
Sonoma County Association of Realtors	
CA—East Bay Regional Data (EBRD)	San Francisco East Bay
Alameda Association of Realtors	
Berkeley Association of Realtors	
Delta Association of Realtor	
Oakland Association of Realtors	
West Contra Costa Association of Realtors	
CA—Fresno Association of Realtors (Fresno)	Fresno California Area
Fresno Association of Realtors	

CA—Lake County Board of Realtors (LCBOR)	Lake County California
Lake County Board of Realtors	
South Lake Tahoe Board of Realtors	
CA—MAX MLS (MAX)	San Francisco East Bay
Bay East Association of Realtors	
Contra Costa Association of Realtors	
CA—MetroList Services, California (Metrolist)	Sacramento/Central Valley California
Central Valley Association of Realtors	
El Dorado County Association of Realtors	
Lodi Association of Realtors	
Merced County Association of Realtors	
Placer County Association of Realtors	
Sacramento Association of Realtors	
Yolo County Board of Realtors	
CA—RE InfoLink (ReInfoLink)	San Francisco South Bay/Peninsula
Monterey County Association of Realtors	
San Benito County Association of Realtors	
San Mateo County Association of Realtors	
Santa Clara County Association of Realtors	
Santa Cruz Association of Realtors	
Silicon Valley Association of Realtors	
Watsonville Association of Realtors	
CA—San Francisco Association of Realtors (SFAR)	San Francisco County California
San Francisco Association of Realtors	
CA—Shasta Association of Realtors (SHASTA)	Shasta
CA—Tuolumne County Association of Realtors (TCAR)	Tuolumne County California
Tuolumne County Association of Realtors	
SOUTHERN CALIFORNIA	
CA—Central Coast MLS (CCR MLS)	California Central Coast
Atascadero Association of Realtors	
Pismo Coast Association of Realtors	
San Luis Obispo Association of Realtors	

CA—Conejo Valley MLS (CVAR)	Conejo Valley CA Area
CA—CRISNET Regional MLS (CRISNET)	San Fernando Valley California
Burbank Association of Realtors	
Conejo Valley Association of Realtors	
Simi Valley-Moorpark Association of Realtors	
Southland Regional Association of Realtors	
CA—Desert Area MLS (Desert MLS)	Palm Springs Desert Area
California Desert Association of Realtors	
Desert Communities Association of Realtors	
Palm Springs Board of Realtors	
CA—Greater South Bay Regional MLS (GSBR)	Long Beach/Southern LA County
Inglewood Board of Realtors	
Long Beach-Southern LA County	
Palos Verdes Peninsula Association of Realtors	
South Bay Association of Realtors	
CA—MRMLS (MRMLS)	Riverside/Fullerton California Area
Arcadia Association of Realtors	
Citrus Valley Association of Realtors	
Corona-Norco Association of Realtors	
Inland Valleys Association of Realtors	
Montebello District Board of Realtors	
Redlands Association of Realtors	
Riverside/Moreno Valley Area Association of Realtors	
Southwest Riverside County Association of Realtors	
Tri-Counties Association of Realtors	
West San Gabriel Valley Association of Realtors	
Yucaipa Valley Board of Realtors	
CA—Pasadena-Foothills Association of Realtors (PFAR)	Pasadena/Glendale California Area
Glendale Association of Realtors	
Pasadena-Foothills Association of Realtors	

CA—Sandicor, Inc. (SANDICOR)	San Diego County California
Coronado Association of Realtors	
East San Diego County Association of Realtors	
North San Diego County Association of Realtors	
Pacific Southwest Association of Realtors	
San Diego Association of Realtors	
CA—Santa Barbara Assoc. of Realtors (SBAOR)	Santa Barbara CA Area
Santa Barbara Association of Realtors	
CA—Southern California MLS (SOCAL)	Orange County California
Downey Association of Realtors	
Laguna Board of Realtors	
Newport Beach Association of Realtors	
Orange Coast Association of Realtors	
Orange County Association of Realtors	
Pacific West Association of Realtors	
Rancho Southeast Association of Realtors	
CA—Ventura County MLS (VC MLS)	Ventura Country CA
Ventura County Coastal Association of Realtors	
PACIFIC NORTHWEST	
ID—Intermountain MLS (IMLS)	Idaho Intermountain
Ada County Association of REALTORS	
Elmore County Board of Realtors	
Emmett Valley Association of REALTORS	
Greater Twin Falls Association of Realtors	
Idaho Association of REALTORS	
Malheur County Board of Realtors	
Nampa Association of Realtors	
North Side Board of Realtors	
Payette Washington Board of Realtors	

OR—Regional Multiple Listing Service, Oregon (RMLS)	Portland OR Area
Baker County Board of Realtors	
Central Oregon Association of Realtors	
Central Oregon Coast Board of Realtors	
Columbia Basin Board of Realtors	
Columbia County Board of Realtors	
Coos County Board of Realtors	
Cottage Grove Board of Realtors	
Curry County Board of Realtors	
Douglas County Board of Realtors	
E. Metro Association of Realtors	
Eugene Association of Realtors	
Grants Pass Association of Realtors	
Klamath County Association of Realtors	
Mid Columbia Association of Realtors	
N. Willamette Association of Realtors	
Polk County Association of Realtors	
Portland Metro Association of Realtors	
Rogue Valley Association of Realtors	
Salem Association of Realtors	
Santiam Board of Realtors	
Springfield Board of Realtors	
The Coos County Board of Realtors	
Tillamook County Board of Realtors	
Umatilla County Board of Realtors	
Union County Board of Realtors	
Wallowa County Board of Realtors	
Willamette Association of Realtors	
Yamhill County Board of Realtors	
OR—Southern Oregon MLS (SO MLS)	**Southern Oregon**
WA—Northwest Multiple Listing Service (NWMLS)	**Seattle Washington Area**
Clark County Association of Realtors	
Cowlitz County Association of Realtors	

Grant County Association of Realtors	
Grays Harbor Association of Realtors	
Jefferson County Association of Realtors	
Kitsap County Association of Realtors	
Kittitas County Association of Realtors	
Lewis County Association of Realtors	
Lewis-Clark Association of REALTORS	
Lower Yakima Valley Association of Realtors	
Mason County Association of Realtors	
Moses Lake Othello Board of Realtors	
North Central Washington Association of Realtors	
North Puget Sound Association of Realtors	
Northeast Washington Association of Realtors	
Orcas Island Association of Realtors	
Port Angeles Association of Realtors	
San Juan County Association of Realtors	
Seattle King County Association of Realtors	
Sequim Association of Realtors	
Snohomish County Camano Association of Realtors	
Spokane Association of Realtors	
Tacoma—Pierce County Association of Realtors	
Tri City Association of Realtors	
Walla Walla Board of Realtors	
Whatcom County Association of Realtors	
Whidbey Island South Association of Realtors	
Whitman County Association of Realtors	
Yakima Association of Realtors	
HI—Honolulu Board of Realtors (HBR)	**Honolulu Hawaii Area**
Honolulu Board of Realtors	

SOUTHWEST	
AZ—Tucson Association of Realtors (TAR)	Tucson Arizona Area
Tucson Association of Realtors	
NM—Southwest Multiple Listing Service (NMSWMLS)	Albuquerque
Tucson Association of Realtors	
NV—Greater Las Vegas Association of Realtors (GLVAR)	Las Vegas Nevada Area
Greater Las Vegas Association of Realtors	
TX—Austin Board of Realtors/ACTRIS MLS (ACTRIS)	Austin Texas Area
Austin Board of Realtors	
TX—North Texas Real Estate Information Systems (NTREIS)	Dallas TX Area
Collin County Association of Realtors	
Ellis Hill County Association of Realtors	
Granbury Board of Realtors	
Grand Prairie Board of Realtors	
Greater Dallas Association of Realtors	
Greater Denton/Wise County Association of Realtors	
Greater Fort Worth Association of Realtors	
Greater Lewisville Association of Realtors	
Greater Texoma Association of Realtors	
Irving-Las Colinas Association of Realtors	
Johnson County Association of Realtors	
Kaufman-Van Zandt Board of Realtors	
Lake Cities Association of Realtors	
Navarro County Board of Realtors	
Northeast Tarrant County Board of Realtors	
Stephenville Association of Realtors	
Weatherford Parker Association of Realtors	

TX—San Antonio Board of Realtors (SABOR)	San Antonio Texas Area
San Antonio Board of Realtors	
COLORADO	
CO—Information and Real Estate Services, LLC (IRES)	Northern Colorado
Boulder Area Board of Realtors	
Estes Park Board of Realtors	
Fort Collins Board of Realtors	
Greeley Area Association of Realtors	
Longmont Association of Realtors	
Loveland/ Berthoud Association of Realtors	
Morgan County Board of Realtors	
Northern Colorado Commercial Association	
CO—MetroList MLS, Colorado (Metrolist MLS)	Denver Colorado Area
Denver Board of Realtors	
Denver Metropolitan Commercial Association of Realtors	
North Metro Denver REALTOR Association	
South Metro Denver Realtor Association	
CO—Pikes Peak MLS (PP MLS)	Pikes Peak Colorado Area
Pikes Peak Association of Realtors	
MIDWEST	
IL—Multiple Listing Service of Northern Illinois (MLSNI)	Chicago Illinois Area
Aurora Tri-County Association of Realtors	
Chicago Association of Realtors	
Fox Valley Association of Realtors	
Illinois Association of Realtors	
Lake County Association of Realtors	
McHenry County Association of Realtors	
North Shore-Barrington Association of Realtors	
Oak Park Board of Realtors	
Realtor Association of Northwest Chicagoland	

REALTOR Association of West/South Suburban Chicagoland	
Three Rivers Association of Realtors	
West Towns Board of Realtors	
Will-Grundy Association of Realtors	
IL—Southwestern Illinois MLS (SIR MLS)	**Southwestern Illinois**
KS—Heartland MLS (HMLS)	**Heartland**
Heartland MLS	
Johnson County Board of Realtors	
Kansas City Regional Association of Realtors	
MI—Grand Rapids Assoc of Realtors (GRAR)	**Grand Rapids MI Area**
Grand Rapids Association of Realtors	
MI—RealComp II (REALCOMP)	**Detroit MI Area**
Dearborn Board of Realtors	
Detroit Area Commercial Board of Realtors	
Livingston Association of Realtors	
Metropolitan Consolidated Association of Realtors	
North Oakland County Association of Realtors	
Western Wayne Oakland County Association of Realtors	
MN—Regional MLS of Minnesota (Regional MLS)	**Minneapolis-St.Paul MN Area**
Minneapolis Area Association of Realtors	
North Metro Realtors Association	
Southern Twin Cities Association of Realtors	
St. Paul Area Association of Realtors	
Western Wisconsin Realtors Association	
OH—Columbus Board of Realtors (CBR) (CBR)	**Columbus Ohio Area**
Columbus Board of Realtors	
OH—Dayton Area Board of Realtors (DABR)	**Dayton Ohio Area**
Dayton Area Board of Realtors	

TN—Knoxville Area Association of REALTORS (KAAR)	Knoxville TN Area
TN—RealTracs (RealTracs)	Mid Tennessee Area
Clarksville Association of Realtors	
Eastern Middle TN Association of Realtors	
Greater Nashville Association of Realtors	
Middle Tennessee Association of Realtors	
Robertson County Association of Realtors	
Rutherford County Association of Realtors	
Southern Middle Tennessee Assn. of Realtors	
Sumner Association of Realtors	
Warren County Board of Realtors	
Williamson County Association of Realtors	
Wilson County Association of Realtors	
NEW ENGLAND	
CT—Consolidated MLS, Inc. (CMLS)	Western Connecticut
Bridgeport Association of Realtors	
Greater Fairfield Board of Realtors	
Greater Hartford Association of Realtors	
Mid-Fairfield County Association of Realtors	
New Canaan Board of Realtors	
New Milford Board of Realtors	
Northern Fairfield County Association of Realtors	
Ridgefield Board of Realtors	
Stamford Board of Realtors	
Valley Association of Realtors	
CT—Darien MLS (Darien MLS)	Darien CT Area
Darien Board of Realtors	
CT—Eastern CT REALTORS® Information Service (ECRIS)	Eastern Connecticut
Eastern Connecticut Association of Realtors	

CT—Greater New Haven/CO-OP MLS (CO-OP MLS)	**Central Connecticut**
Greater New Haven Association of Realtors	
Greater Waterbury Board of Realtors	
Litchfield County Board of Realtors	
Middlesex/Shoreline Association of Realtors	
CT—Greenwich MLS (Greenwich)	**Greenwich CT area**
Greenwich Association of Realtors	
CT – New Canaan MLS (New Canaan)	**New Canaan CT Area**
MA—Cape Cod & Islands MLS, Inc (CCI MLS)	**Cape Cod and Islands— MA**
MA—Massachusetts (MLSPIN)	**Massachusetts**
Berkshire County Board of Realtors	
Eastern Middlesex Association of Realtors	
Greater Boston Real Estate Board	
Greater Fall River Association of Realtors	
Greater New Bedford Association of Realtors	
Greater Newburyport Association of Realtors	
North Central Mass. Association of Realtors	
North Shore Association of Realtors	
Northeast Association of Realtors	
Plymouth and South Shore Association of Realtors	
Realtor Association of Pioneer Coun ty	
Tri-County Association of Realtors	
Worcester Regional Association of Realtors	
ME—Maine Real Estate Information System (MREIS)	**Maine**
Androscoggin Valley Board of Realtors	
Bangor Board of Realtors	
Hancock-Washington Board of Realtors	
Kennebec Valley Board of Realtors	
Lincoln County Board of Realtors	
Maine Association of Realtors	
Maine Commercial Association of Realtors	
Maine Living Network	

Merrymeeting Board of Realtors	
Mid-Coast Board of Realtors	
Mountains Council of REALTORS®	
Portland Board of Realtors	
Western Maine Board of REALTORS®	
York County Council of Realtors	
NH—Northern New England Real Estate Network (NNEREN)	**Northern New England— NH & VT**
Addison County Board of Realtors	
Central Vermont Board of Realtors	
Concord Board of Realtors	
Contoocook Valley Board of Realtors	
Crown Point Board of Realtors	
Franklin County Vermont Board of Realtors	
Granite State South Board of Realtors	
Greater Claremont Board of Realtors	
Greater Manchester Nashua Board of Realtors	
Lakes Region Board of Realtors	
Lamoille Area Board of Realtors	
Monadnock Region Board of Realtors	
New Hampshire Association of Exclusive Buyer Agent	
New Hampshire Association of Realtors	
North Country Board of Realtors	
Northeast Kingdom Board of Realtors	
Northwestern Vermont Board of Realtors	
Orleans County Board of Realtors	
Rutland County Board of Realtors	
Seacoast Board of Realtors	
South Central Vermont Board of Realtors	
Southeastern Vermont Board of Realtors	
Southwestern Vermont Board of Realtors	

Strafford County Board of Realtors	
Sunapee Region Board of Realtors	
Upper Valley Board of Realtors	
White Mountain Board of Realtors	
Windsor County Board of Realtors	
NY—Greater Hudson Valley MLS (GHV MLS)	Hudson Valley New York Area
Orange County Association of Realtors	
Rockland County Board of Realtors	
NY—Mid-Hudson MLS (Mid-Hudson MLS)	Mid-Hudson NY Area
Dutchess County Association of Realtors	
NY—Westchester-Putnam Mulitple Listing Service (WP MLS)	Westchester -Putnam NY
Putnam County Board of Realtors	
Westchester County Board of Realtors	
RI—Rhode Island State Wide MLS (Statewide MLS)	Rhode Island
Greater Providence Board of Realtors	
Kent Washington Board of Realtors	
Newport County Board of Realtors	
Northern Rhode Island Board of Realtors	
Rhode Island Association of Realtors	
Rhode Island Commercial And Appraisal Board of Realtors	
CENTRAL EAST COAST	
DC—METROPOLITAN REGIONAL INFORMATION SYSTEMS (MRIS)	Maryland-Washington DC Area
Anne Arundel County Association of Realtors	
Bay Area Association of Realtors	
Blue Ridge Association of Realtors	
Carroll County Association of Realtors	
Cecil County Board of Realtors	
Dulles Area Association of Realtors	
Eastern Panhandle Board of Realtors	

Frederick County Association of Realtors	
Fredericksburg Area Association of Realtors	
Garrett County Board of Realtors	
Greater Baltimore Board of Realtors	
Greater Capital Area Association of Realtors	
Greater Piedmont Area Association of Realtors	
Harford County Association of Realtors	
Historic Highlands Association of Realtors	
Howard County Association of Realtors	
Maryland Association of Realtors	
Massanutten Association of Realtors	
Mid-Shore Board of Realtors	
Northern Virginia Association of Realtors	
Pen-Mar Regional Association of Realtors	
Potomac Highlands Board of Realtors	
Prince George's County Association of Realtors	
Prince William Association of Realtors	
Southern Maryland Association of Realtors	
Virginia Association of Realtors	
Washington DC Association of Realtors	
NC—Carolina MLS (Carolina MLS)	**Charlotte North Carolina Area**
Central Carolina Association of Realtors	
Charlotte Region Commercial Board of Realtors	
Charlotte Regional Realtor Association	
Gaston Association of Realtors	
Union County Association of Realtors	
NC—Outer Banks Association of REALTORS® (OBAR)	**North Carolina Outer Banks**
Outer Banks Association of Realtors	
NC—Triangle MLS (TMLS)	**Greater Raleigh NC Area**
Chapel Hill Board of Realtors	
Durham Association of Realtors	

Raleigh Regional Association of Realtors	
Raleigh/Wake Board of Realtors	
NC—Wilmington Regional Assoc of Realtors (WRAR)	**Wilmington**
Brunswick County Board of Realtors	
Wilmington Regional Assoc of Realtors	
NJ—Garden State MLS (GS MLS)	**Northern New Jersey Area**
Eastern Bergen County Board of Realtors	
Greater Union County Association of Realtors	
Hunterdon—Somerset Association of Realtors	
Meadowlands Board of Realtors	
Middlesex County Association of Realtors	
North Central Jersey Association of Realtors	
Passaic County Board of Realtors	
Real Source Association of Realtors	
Sussex County Association of Realtors	
Tri-State Commercial & Industrial Association of Realtors	
United Association of Realtors	
Warren County Board of Realtors	
West Bergen—RealSource Association of Realtors	
West Essex Board of Realtors	
NJ—Monmouth/Ocean MLS (MO MLS)	**Monmouth County NJ Area**
Monmouth County Association of Realtors	
Ocean County Board of Realtors	
South Monmouth Board of Realtors	
NJ—Ocean County Board of Realtors (OCBR)	**Ocean County NJ Area**
Ocean City Board of Realtors	
PA – Trend MLS (TReND)	**Philadelphia PA Area**
Burlington Camden County Association of Realtors	
Gloucester Salem Counties Board of Realtors	
Kent County Association of Realtors	
Mercer County Association of Realtors	
New Castle County Board of Realtors	

SC—Coastal Carolinas Association of REALTORS®, Inc. (CCAR)	Coastal South Carolina Area
Coastal Carolinas Association of Realtors	
Grand Strand Board of Realtors	
VA—Central Virginia Regional MLS (CVR MLS)	Richmond VA Area
Richmond Association of Realtors	
Southside Virginia Association of Realtors	
SOUTH	
AL—Birmingham Association of REALTORS (BHAM)	Birmingham AL Area
Birmingham Association of Realtors	
FL—Bonita Springs MLS (Bonita Springs MLS)	Bonita Springs FL Area
FL—Cape Coral MLS (Cape Coral MLS)	Cape Coral FL Area
Cape Coral Association of Realtors	
FL—Chipola Area Board of Realtors (Chipola)	Chipola FL Area
FL—Englewood MLS (Englewood)	Englewood FL Area
Englewood Area Board of Realtors	
FL—Florida Keys Marathon MLS (Florida Keys)	Florida Keys
Florida Keys Board of Realtors	
Key West Association of Realtors	
Marathon and Lower Keys Association of Realtors	
FL—Fort Lauderdale MLS (Ft Lauderdale)	Fort Lauderdale FL Area
Realtor Association of Greater Fort Lauderdale	
FL—Greater Ft. Myers and the Beach (Ft. Myers)	Ft. Myers FL Area
FL—Miami MLS (Miami)	Miami FL Area
Northwestern Dade Association of Realtors	
Realtor Association of Dade County	
Realtor Association of Greater Miami and the Beach	
Realtor Association of Miami—Dade County	
South Broward Association of Realtors	
FL—Mid-Florida Regional MLS (MFR)	Orlando/Tampa FL Area
Bartow Association of Realtors	
East Polk County Association of Realtors	

Greater Lake County Association of Realtors	
Greater Orlando Association of Realtors	
Greater Tampa Association of Realtors	
Lakeland Association of Realtors	
Orlando Regional Realtor Association	
Osceola County Association of Realtors	
Pinellas Realtor Organization	
FL—Northeast Florida MLS, Inc. (NE Florida MLS)	**Jacksonville FL Area**
Northeast Florida Association of Realtors	
St. Augustine & St. Johns County Board of Realtors	
FL—Ocala Marion Board of Realtors (OMCBR)	**Ocala-Marion FL Area**
Ocala / Marion County Association of Realtors	
FL—Punta Gorda/Port Charlotte (Punta Gorda)	**Punta Gorda/Pt. Charlotte Area**
Manatee Association of Realtors	
FL—Regional Multiple Listing Service, Florida (Regional MLS)	
Boca Raton & Highland Beach Association	
Boyton Beach & Del Ray Beach Association	
Broward County Association of Realtors	
Commercial Society of Palm Beach County/ Boca Raton	
Jupiter/ Tequesta/Hobe Sound Association of Realtors	
Martin County Association of Realtors	
Palm Beaches Realtor Association	
Realtor Association of the Palm Beaches	
St. Lucie Association of Realtors	
FL—Sarasota MLS (Sarasota MLS)	**Sarasota FL Area**
Sarasota Association of Realtors	
FL—Spacecoast and Melbourne MLS (Space Coast)	**Florida Space Coast Area**
Melbourne Area Association of Realtors	
Space Coast Association of Realtors	

GA—Columbus Board of Realtors MLS (CBOR)	Columbus Georgia area
Columbus Board of Realtors	
GA—First Multiple Listing Service, Inc. (First MLS)	Georgia
Albany Board of Realtors	
Athens Area Association of Realtors	
Atlanta Board of Realtors	
Brunswick-Glynn County Board of Realtors	
Carpet Capital Association of Realtors	
Cherokee Association of Realtors	
Cobb Association of Realtors	
Dekalb Board of Realtors (Metro Atlanta)	
Dublin Board of Realtors	
Fayette County Board of Realtors	
Georgia Association of Realtors	
Greater Augusta Association of Realtors	
Greater Capital Area Association of Realtors	
Greater Rome Board of Realtors	
Hall County Board of Realtors	
Heart of Georgia Board of Realtors	
I-85 North Board of Realtors	
Lake Oconee Area Association of Realtors	
Metro South Association of Realtors	
Middle Georgia Commercial Council	
Newnan-Coweta Board of Realtors	
Northeast Atlanta Metro Association of Realtors	
Northeast Georgia Board of Realtors	
Savannah Board of Realtors	
Southeast Georgia Board of Realtors	
West Georgia Board of Realtors	
West Metro Board of Realtors	
GA—Georgia MLS (Georgia MLS)	Georgia
Atlanta Commercial Board of Realtors	
LA—Greater Baton Rouge MLS (GBRAR MLS)	Baton Rouge LA Area

Index

123AnySt.com, 6, 7, 50, 81, 82, 85, 86, 89, 142, 165, 213

24x7, 8, 13, 21, 46, 55, 169

360-degree, 106
3rd party websites, 21, 133

56k modem, 54, 109

Act!, 67
Activity, 143, 207
Adobe Acrobat, 229
Adobe Photoshop, 230
Advertising, 24, 78, 83, 85, 97, 251
Advertising coordinator, 24
AdWords, 78
Affiliate, 31
Age, 22
Age Gap, 22
Agency ABC, 201
AgencyLogic, ii, xviii, xx, 7, 49, 50, 82, 89, 98, 99, 100, 115, 142, 165, 213
Agent account, 161
Agent websites, 5

AgentAVM, 49, 158, 159, 160, 163, 164, 165, 166
Agriculture, 92
Alt tags, 62
Application, 46, 203
Appointments, 250, 284
Appraisal, 309
Arizona Regional Multiple Listing Service, 128
ARMLS, 128
Art, 148, 230, 232
Assessment, xix
Assistants, xviii, 5, 16, 22, 27, 32, 34, 62, 194, 209, 249, 252, 266, 267, 271, 291
Associations, 170, 189
Atomic Energy Complex, 145
Audience, 4, 153
Automated, xviii, 5, 11, 21, 23, 152, 153, 239
AvalancheRealEstate.com, 80
Average, 227
AVM, xviii, 3, 5, 11, 12, 21, 23, 45, 46, 47, 48, 49, 123, 147, 152, 153, 154, 155, 156, 157, 158, 159, 160, 161, 162, 163, 165, 239
Awareness, 22

BlackBerry, 219
Blogs, 57
Bold text, 76

Book, xxvii, 290
BPO (Freddie Mac), 172
Broadband, 54
Brochure, 69, 100
Browser, 59, 60, 61, 62, 64

Cable, 109, 182, 226
Cache, 226
Calculators, 46, 50, 223
calendar, ii, 218, 220
Casino, 95
CD, 9, 113, 199, 200, 230, 231, 232
Census tracts, 144
Center for Realtor Technology, 202
Certificants, 241
Certification, xix, xxvii, 241, 242
CERTIFIED FINANCIAL PLANNER, 30
Charts, 153
Checklist, 70
CitySearch, 78
Clareity Consulting, 26, 179, 181, 186, 187, 191
Clark University's Graduate School of Geography, 145
Clerical, 283, 284, 291
CMA, 5, 7, 10, 11, 12, 33, 45, 46, 47, 48, 49, 123, 147, 148, 149, 150, 151, 152, 153, 155, 157, 158, 161, 162, 166, 229, 239, 256
CMA statistics, 149
CMA-looking, 11, 12, 166
CNN.com, 78
CNNfn, 78
Commercial real estate, 89
Commission, 246, 247, 248, 249, 252, 253, 254, 255, 257, 258, 260, 264, 265, 266, 267, 268, 270, 271, 273, 274, 276, 278, 280, 281, 282, 283,

285, 286, 287, 288, 289, 290, 291, 292, 293, 294
Communication, 32, 186, 191, 203, 204, 205
Communities, 46, 136, 299
Comparable Market Analysis, 5, 10, 147, 239
Consumer, 154, 155, 190, 257
Contact forms, 6
Contact information, 71, 120
Contact management, 36
Contact Relationship Management, 67, 187
Corporate Sponsor, xix, 6, 7, 8, 10, 12, 13, 14, 16, 17, 18, 31, 47, 67, 68, 239
County, 153, 291, 297, 298, 299, 300, 301, 302, 303, 304, 305, 306, 307, 308, 309, 310, 311, 312, 313, 314
Course, 95
CRM, 6, 13, 17, 67, 179, 183, 187, 204, 222
CRT, xxi, 22, 126, 149, 158, 178, 186, 188, 192, 202
Customers, 19
Cybertrust, 203

Data Accuracy, 149, 157
Databases, 57, 153
DBA, 64
Deposit, 284
Description tag, 76
Designation, 71
Desktop Application, 173
Dial-up, 54
Digital signatures, 198
Directional Pad, 219
Discussion, 241
Document Viewing, 223
Documents, 46, 50, 100, 251, 284, 285

Domain names, 44, 49
Drip marketing, 36
Duties, 284
DVD/CD-RW, 226

eAppraiseIt, iii, xvii, 12, 156, 158, 159, 160, 165
Edit, 114, 115, 121
Electricity, 182
Electronic, ii, xvii, xxiii, 5, 12, 13, 21, 45, 47, 49, 167, 169, 170, 172, 173, 176, 178, 179, 181, 198, 238, 239
Electronic Forms, xvii, xxiii, 5, 12, 21, 45, 47, 49, 167, 169, 170, 172, 179, 181, 198, 238, 239
Electronic forms software, 172, 178
Electronic Signatures, 173, 176, 198
Electronic Signatures in Global and National Commerce Act (ESIGN), 176
Electronic transactions, 12, 21, 169
Email, 44, 57, 66, 67, 68, 69, 100, 120, 134, 204, 207, 219, 240
Employee Relocation Council (ERC), 172
Employment, 278
Entrepreneurial Interests, xxviii
e-Pro, xx
Escrow, 207, 284
ESIGN, 176
Etiquette, 64
Evolution, 145
Excel, 148, 151, 223
Executive, i, ii, iii, xvii, xviii, xix, 90, 136
Exposure settings, 227

Families, 283
FAQ, 70, 232
Fax, 127, 195, 196
FBI, 10, 135, 143

Feature, 69, 100, 101, 120, 121, 122, 284
File Transfer Protocal, 232
Financial Services, 182
Flash, 57, 71, 227
Flash Media Card, 227
Flexibility, 150, 157
Footnotes, xxix
FormMail, 57
Forms, 13, 45, 49, 62, 70, 71, 134, 183, 188, 198, 199, 284, 285
Forums, 26, 241
Frame
Framed IDX, 128
FSBO, 23, 87, 88, 154
FTP, 57, 127, 129, 130, 232, 233
Funds, 251

Geography, 145
Google, 69, 78, 79
Graphs, 153
Guestbooks, 57
Guide, 105, 243, 245

H tag, 76
Handbook, vii, viii
Hardware, xxviii, 215, 225
HDVT, 8, 119, 122
High Speed Internet, 226
History, 145, 205
Home buyers, 8, 9
Homegain, 78
HomeGain, 135
HomeTour360, 118
Hospitality, 90, 92, 95
Hosting, 56, 57, 58, 59, 62, 64, 66, 99, 128, 134, 232
Hosting companies, 58, 66

Hotel, 95
HTML, xxvi, 6, 46, 59, 60, 68, 76, 98, 99

JPEG, 109, 226, 227, 231

IDX, 5, 6, 45, 46, 48, 71, 78, 111, 119,
123, 125, 126, 127, 128, 129, 130,
131, 239, 243, 297
ILD, 5, 123, 125, 131, 239
Image format, 227
Images, 121, 227
Information Sciences, 145
Information Technology Association of
America, 31
Ingenio, 80
Inspect, 284
Institutions, 153
Integration, 5, 13, 71, 169, 178, 179, 198,
239
Internet, xx, xxiii, xxiv, xxvi, xxvii, xxviii,
xxxi, 4, 7, 13, 19, 23, 41, 43, 53, 54,
55, 59, 60, 61, 62, 64, 66, 73, 75, 77,
79, 86, 97, 99, 125, 126, 127, 130,
131, 133, 154, 174, 177, 182, 183,
185, 187, 190, 200, 202, 218, 286
Internet access, xxiv, 53, 55, 59, 60, 86,
97, 99, 126, 174, 177, 218
Internet applications, xxvii, 53
Internet connection, 13, 53, 54, 61, 183
Internet connection speeds, 54
Internet Crusade, xx
Internet Data eXchange, 125, 127
Internet Service Provider, 53, 55, 66
Internet-enabled PC, 175, 185, 190
Intimidation, 22
IRS Change of Address, 182
iseemedia, ii, 16, 120, 220, 221, 222
iseerealty, 16, 50, 220
ISP, 54, 55, 56, 66
IT, 29, 31, 230

Keywords, 76
Kitchen, 86

Landlord, 172
Lead forms, 46, 48
Lead-based Paint Addendum, 172
Lead-generation, 142, 154
Leads, 134
Learning Laboratory, xix, 17, 18
Leasers, 89
Lenders, 284
Lens, 227
Licensee, 277, 279, 280, 284, 285
Link Exchange, 80
LinkRE.com, 80
Linux, 57
Local Match, 77, 78
Location Inc, ii, 144
Logging, 186, 191, 203, 205, 206
Logo, 101
Loopnet.com, 84, 98

Macromedia Fireworks, 230
Main Internet Presence, 53
Maintenance, 33
Management, 14, 57, 88, 136, 179, 185,
186, 188, 191, 192, 194, 214
Mapping, 69
Maps, 46, 50, 69
Market, 30, 150, 154, 250, 276, 283
Marketing, xvii, xxx, 7, 13, 33, 73, 101,
271
Medical, 89, 90

Message Boards, 57

Messages, 284

Meta-tag, 77

Microsoft Outlook, 13, 67, 179, 183, 187, 204

Microsoft Word, 194

MLS, ii, xxiv, 5, 6, 11, 12, 13, 15, 16, 18, 20, 24, 25, 33, 42, 43, 44, 45, 46, 48, 62, 71, 78, 88, 111, 119, 123, 125, 126, 127, 128, 129, 130, 131, 133, 147, 148, 149, 150, 151, 152, 153, 157, 161, 162, 169, 170, 172, 174, 178, 179, 183, 189, 197, 198, 210, 211, 217, 220, 221, 222, 224, 239, 243, 273, 277, 292, 297, 298, 299, 300, 301, 303, 304, 305, 306, 307, 309, 310, 311, 312, 313, 314

MLS CMA, 150, 151, 152, 157, 162

Mobile technology, 47

Mobility, ii, xxiv

Model, xviii, 11, 21, 23, 152, 153, 238, 239, 241

Money, 284

Mortgage, 70, 100, 223, 248

Motel, 95

MSN, 78

Multiple Search Engine, 78

MySQL, 57

National Association of Realtors, xx, xxiv, 13, 125, 130, 183

National Center for Education Statistics, 144

National Institute of Webographers, viii, xix, xx, xxi, xxiv, xxv, xxvii, 11, 18, 23, 32, 52, 157, 158, 237, 238, 239, 240, 242

Neighborhood, xviii, xxiii, 5, 9, 45, 46, 47, 48, 49, 93, 123, 132, 134, 135, 136, 137, 138, 139, 140, 144, 145, 162, 239

Neighborhood Search, xviii, xxiii, 5, 9, 45, 46, 47, 48, 49, 123, 132, 134, 135, 137, 138, 139, 140, 162, 239

NeighborhoodScout, ii, xviii, xix, 10, 49, 134, 135, 136, 137, 138, 139, 140, 141, 142, 143, 144, 145, 146, 162, 165

Network, ii, xviii, xx, 7, 134, 146, 226, 307, 308

New Prospect, 33

News, 6, 61, 70, 284

News Feed, 6

Newspaper listings, 20

Newspapers, 84, 85, 96, 97, 182

Nielsen/NetRatings, xxiv

North America, 3, 23, 220, 222

Notebook, 15, 217, 219

Notification, 186, 191, 203, 204

Oak Ridge National Laboratory, 145

Occupational Direction, xxviii, 31

Offers, 285

Office Managers, 24, 26, 98

Office space, 90, 95

One-on-one, 31

Online Transaction Management, xvii, xxiii, 5, 14, 21, 45, 47, 167, 179, 185, 186, 187, 189, 191, 195, 197, 202, 209, 210, 214, 238, 239

Optical zoom, 226

OTM, 14, 21, 45, 47, 179, 180, 181, 182, 185, 187, 189, 190, 191, 192, 194, 195, 196, 197, 198, 199, 200, 201, 202, 203, 204, 205, 206, 208, 209, 210, 211, 212, 213, 214

OTM fax services, 196

Outlet, 93

Outputting, 104
Overture, 77, 78, 79
Owners.com, 87

palmOne Treo, 219
Panorama, 228
Panorama Head, 228
Paradise Valley, 136
Participant Set-Up, 186, 191, 200
PC, 15, 61, 174, 177, 217, 218, 219, 225, 229, 232
PDA, 15, 43, 56, 172, 205, 217, 218, 219, 220
PDF, 13, 69, 170, 194, 229
Permissions, 14, 22
Personal branding, 210
Personalization, 150, 153, 157
Personnel, xix, 285
Phase, 42, 44, 47, 48, 51, 212
Phone call, 67
Photos, 33, 45, 46, 48, 49, 105, 107, 108, 226
Photo-taking, 104, 107
Photovista, 8, 108, 119
PHP, 57
PicturePath, 8, 47, 50, 111, 117, 118, 119
Platform, 57
Pool, 86
POP3, 57
Portrait mode, 109
PowerSite, 7, 89, 99, 115, 142, 165, 213
Price, 55, 71, 89, 271
PrimoPDF, 229
Print, 100
Process, 25, 41, 42, 113, 148, 277, 284
Processor, 226
Professional-types, 136
Public Information Records, 153

public record, 35, 153, 247, 275, 279, 291, 294
Publish, 107, 110, 111, 120, 121

Questions, 42, 43, 269, 283, 285

RAM, 226
Rapidlistings, 69, 134
Real Estate Business Technologies, xvii, xix, 186, 198, 202, 204, 206, 209, 214
Real Estate FAQs, 6
Real estate professionals, i, 178, 199
Real Estate Transaction Standard, 126, 178
RealBiz360 Mobile, 224
Realbiz360.com, 8, 113, 117, 119
RealEstateLinkExchange.com, 80
REALTOR Secure, 202
Realtor.com, 8, 9, 21, 27, 44, 46, 47, 50, 84, 102, 111, 116, 117, 118, 119, 178, 237
Refer a Friend, 45, 62
Region, 308, 309, 310
RELAY, xvii, xix, 14, 17, 35, 49, 180, 181, 183, 190, 191, 192, 194, 195, 196, 197, 198, 199, 200, 201, 202, 203, 204, 205, 206, 207, 208, 209, 210, 211, 212, 213, 214
Rental, 71, 95, 128, 284, 285
Report, xviii, 47
Requirements, 42, 43, 44
Research, 90, 91, 239, 275
Resolution, 109, 226
Resort, 92, 95
Resources, xxviii, 21, 58, 70, 219
Resumé, 37, 38
Retail, 90, 92, 94, 95
RETS, 126, 178, 179

Role, 27, 186
Royalty-Free, 231

Sale, 87, 89, 91, 93, 94, 128, 284
Scanner, 228
Science for The Nature Conservancy, 145
Scripts, 57
Searching & Reporting, 207
Secure, 57, 122, 256, 259, 262, 264, 266,
    267, 272, 274, 280, 281, 288, 289,
    292
Secure Shell, 57
Secure Socket Layer, 57
Security, 57, 186, 191, 200, 202, 203, 285
Self-managed, 36, 245
Self-serve, 104
Sellers, xxiv, 8, 14, 21, 68, 84, 98, 283
Selling Mistakes, 70
Service Ordering, 186, 191, 207
service providers, xxv, 3, 18, 27, 32, 41,
    47, 48, 51, 53, 113, 118, 181, 182,
    198, 218, 238, 251
Setup, 71
Shockwave, 57
Shopping, 90, 93
Showcase
Signage, 65, 86
Signatures, 176, 177, 198
Single-property websites, 6, 45, 49, 81, 83,
    88, 89, 98
Site visitors, 130, 133
SmartPhone, 218, 220
SMTP, 57
Software, 13, 57, 169, 170, 183, 215, 225,
    227, 229, 232
Solicit, 35, 269, 288
Solutions, 120
Spa, 95
Specials, 61

Speeds, 54
SSL, 57
Staffing, xviii, 186
Staged, 207
Standards, 270
Stitching, 104, 107, 120, 121
Stock Illustrations, 232
Stock Photography, 230, 231, 232
Stock Video, 232
Subscription, 140
Support staff, 24, 27

T1, 54, 109, 226
Table, ix, 62
Tasks, 192
TCRE, 17, 246
TCTC, 17, 35
Team Contractor Real Estate, 17, 246
Team Contractor Transaction Coordinator,
    17
Team Double-Click, iii, xviii, 17, 32, 34,
    35, 36, 50, 186, 201, 209, 245, 246
Technology, i, vii, viii, xx, xxi, 1, 3, 5, 6,
    16, 22, 23, 25, 26, 27, 50, 52, 54, 67,
    75, 126, 129, 143, 149, 158, 173,
    178, 186, 188, 189, 192, 215, 217,
    220, 235, 238, 239, 240, 241
Telephone, 182, 280, 284
Telnet, 57
Template, 175
TEMPO, 128
Tenants, 91
Testimonial, 70
Thumb drive, 226
Tips, 130
Title tag, 76
Top Producer, 13, 67, 179, 183, 187
TourBuilder, 108, 109, 120, 121
Track, 100

Training, xix, 202, 239, 240, 290
Transaction, 14, 24, 28, 34, 36, 49, 167, 179, 180, 185, 186, 188, 191, 199, 207, 208, 214, 279, 283, 284
Transaction CD, 199
Transaction coordination, 36
Tripod, 107, 227, 228

U

U.S. Bureau of the Census, 144
U.S. Department of Justice, 144
U.S. Geological Service, 144
Unix, 57
Upgrades, 86
URL, 6, 7, 47, 50, 63, 64, 65, 71, 81, 82, 84, 85, 86, 87, 88, 89, 97, 98, 111, 114, 115, 116, 117, 128, 139, 140, 163, 211, 238
USB, 219, 226, 227, 229
User Interfaces, 145

V

VA, xxv, 16, 17, 32, 33, 34, 36, 37, 46, 50, 209, 245, 292, 312
Vacation, 92, 95
Valtira, i, xix, xx, 18
Value-added, 9, 132
Virtual assistant, xxv, 34
Virtual Office Website, 125, 130, 131
Virtual Tour Distribution Network, 116, 117, 118
Virtual tours, 8, 9, 27, 222
VirtualTourWebStore.com, 107, 228
VOW, 5, 123, 125, 130, 131, 239
VTDN, 116, 117, 118, 119

W

Wealthy, 136
Web applications, 3

Web-based technologies, 20
Webographer-friendly hosting company, 53, 59
Webographers.com, xx, xxi, xxvi, xxvii, xxix, 17, 18, 23, 26, 32, 44, 52, 63, 67, 68, 77, 99, 113, 120, 143, 166, 183, 235, 237, 239, 240, 241, 242
Webography, xxv, 3, 5, 32, 39, 41, 42, 44, 47, 48, 51, 52, 53, 57, 58, 113, 180, 198, 212, 225
Webography process, 32, 41, 44, 47, 51, 57, 225
Website statistics, 121
Wide panoramic viewer, 8, 119
Windows, 57, 173, 175, 226, 229
WinForms, 14

Y

Yahoo!, 9, 27, 44, 50, 66, 77, 78, 111, 116, 118
Yahoo! Real Estate, 9, 27, 44, 50, 111, 116, 118

Z

ZipForm, i, xvii, xix, 13, 14, 170, 171, 172, 175, 177, 179, 180, 181, 183, 194, 198, 199, 214
ZipFormConcierge, 181, 182, 184
ZipFormDesktop, 13, 173, 183
ZipFormEsign, 177, 183, 199
ZipFormMLS-Connect, 13, 179, 183, 184, 198
ZipFormMobile, 174, 175
ZipFormOnline, 13, 49, 170, 174, 175, 180, 181, 183, 198
ZipRealty.com, 135
Zoom, 122

978-0-595-39419-7
0-595-39419-1